Probabilistic Aesthetics
of the Avant-Gardes

Technicities

Published

www.edinburghuniversitypress.com/series/TECH

Probabilistic Aesthetics of the Avant-Gardes

Predictive Arts

Konrad Wojnowski

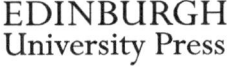
EDINBURGH
University Press

Edinburgh University Press is one of the leading university presses in the UK. We publish academic books and journals in our selected subject areas across the humanities and social sciences, combining cutting-edge scholarship with high editorial and production values to produce academic works of lasting importance. For more information visit our website: edinburghuniversitypress.com

Grateful acknowledgement is made to the sources listed in the List of Illustrations for permission to reproduce material previously published elsewhere. Every effort has been made to trace the copyright holders, but if any have been inadvertently overlooked, the publisher will be pleased to make the necessary arrangements at the first opportunity.

Edinburgh University Press Ltd
13 Infirmary Street
Edinburgh, EH1 1LT

First published in hardback by Edinburgh University Press 2024

Typeset in 11/13 Adobe Sabon by
IDSUK (DataConnection) Ltd

A CIP record for this book is available from the British Library

ISBN 978 1 4744 8896 9 (hardback)
ISBN 978 1 4744 8897 6 (paperback)
ISBN 978 1 4744 8899 0 (webready PDF)
ISBN 978 1 4744 8898 3 (epub)

The right of Konrad Wojnowski to be identified as the author of this work has been asserted in accordance with the Copyright, Designs and Patents Act 1988, and the Copyright and Related Rights Regulations 2003 (SI No. 2498).

Art plays an *unknowing* game with ultimate things,
and yet achieves them!

Paul Klee, *Creative Credo* (1920)

Contents

List of Figures

Series Editors' Preface

Technological transformation has profound and frequently unforeseen influences on art, design and media. At times technology emancipates art and enriches the quality of design. Occasionally it causes acute individual and collective problems of mediated perception. Time after time technological change accomplishes both simultaneously. This new book series explores and reflects philosophically on what new and emerging *technicities* do to our everyday lives and increasingly immaterial technocultural conditions. Moving beyond traditional conceptions of the philosophy of technology and of techne, the series presents new philosophical thinking on how technology constantly alters the essential conditions of beauty, invention and communication. From novel understandings of the world of technicity to new interpretations of aesthetic value, graphics and information, Technicities focuses on the relationships between critical theory and representation, the arts, broadcasting, print, technological genealogies/histories, material culture and digital technologies and our philosophical views of the world of art, design and media.

The series foregrounds contemporary work in art, design and media whilst remaining inclusive, in terms of both philosophical perspectives on technology and interdisciplinary contributions. For a philosophy of technicities is crucial to extant debates over the artistic, inventive and informational aspects of technology. The books in the Technicities series concentrate on present-day and evolving technological advances but visual, design-led and mass mediated questions are emphasised to further our knowledge of their often-combined means of digital transformation.

The editors of Technicities welcome proposals for monographs and well-considered edited collections that establish new paths of investigation.

Ryan Bishop and Jussi Parikka

Chapter I

Introduction[1]

As I am writing this sentence, which – chronologically speaking – happens to be one of the last I intend to write, I am also waiting impatiently for images I have ordered online from an artificial artist, named Midjourney Bot. The program, like many others, currently available to the public through beta testing, relies on a so-called generative adversarial network (GAN), a type of machine learning architecture used to imitate different data distributions, like texts, sounds or images. The most unsettling (and one of the earliest) examples of such imitations are pictures of deceptively realistic human faces of non-existing people, available, for example, at www.this-person-does-not-exist.com. To come up with such images – of people maybe, GANs stage an artificial rivalry between two networks. To come up with such images – of people who never existed, or maybe are yet to be born – GANs stage a computational competition between two semi-independent networks trying to outsmart each other. The first one is called 'discriminator' and was fed in advance with a sample set of human faces labelled as true. The network analyses them and learns to discern human faces from other depictions. Then, the second network, called 'generator', gets to work and tries to fool the first one by feeding it images it created on its own and does so for so long as to be able to consistently convince the discriminator that its portraits are legitimate depictions of human faces; that is, they resemble faces from the discriminator's training set. Crucially, an external source of random noise is introduced so that the generator constantly produces images that look different. Otherwise, the lazy generator would create one convincing image and fool the discriminator by endlessly feeding it with its first 'good-enough' product.

What distinguishes this kind of AI from conventional counterparts, operating on symbolic logic and executing precisely defined

[1] This book is the result of the research project no. 2016/21/D/HS2/02415 funded by the National Science Centre.

operations, is that it does not operate deterministically (the same inputs reliably produce the same outputs) but probabilistically. The hidden model of 'true' faces is a fuzzy approximation of a human visage, dependent on the dataset it was trained on. Furthermore, it remains to some extent indeterminate so it can deal with previously unknown images to analyse and discriminate. Thus, even if it eventually decides that a face is either true or false, it is always an educated guess with a degree of probability (defined arbitrarily as sufficient by its programmers). The same principle applies to the second network that generates images. Its creations are always guesses – blind at first and more educated at the end – and its creativity is bolstered by the addition of a source of random noise. Uncertainty is here not an obstacle to overcome but an inherent component of the whole process that secures the novelty and uniqueness of generated deceptions. For this reason, the generator and the discriminator are often compared to the artist and the critic respectively, which makes the GAN an art-world in a box (of electronics). And more elaborate GANs, like the Midjourney Bot, are capable of generating diverse and ever surprising imagery out of simple commands written in a natural language, like the one which I was impatiently waiting for and which had been intended as a depiction of Albert Einstein and Niels Bohr playing in a colourful ball pit, but somehow ended up as a goofy drawing of an oddly large basket in which contorted faces of ball-shaped physicists lie next to three enormous balls in red and blue. This was, though, a happy disappointment, more felicitous than an accurate representation of my intentions would ever be as I was confronted with something I would never dare to imagine.

Should one take the metaphors of the artist and the critic literally and think of artistic creativity as such as essentially probabilistic, though? Does the same go for art criticism? For some, it is purely a rhetorical question; for others, matters of human creativity and artificial intelligence belong to completely different domains. However, partly to retort to such sceptical traditionalists, I intend to show that the notion of probability – of 'tamed' chance – was already important for many avant-garde artists who denounced old conventions of realist representation and semantic coherence in favour of work methods that involved systematic handling of chance (external randomness generators) and/or embraced uncertainty as a positive force inextricably associated with creativity.[2]

---

[2] During a conference discussion on the perplexing status of art produced by neural networks, which, because of their non-deterministic programming, spark endless debates on the creative aspects of machine intelligences, Jan Løhmann Stephensen asked a fascinating question: Why do most of these images resemble paintings by

My primary assertion is that the probabilistic revolution, while recognised and investigated by historians of science, has been largely overlooked in art. The epistemic turn toward probability has been recognised and described by Ian Hacking as 'the taming of chance' (1990) or, as Gerd Gigerenzer et al. would have it, as an 'empire of chance' taken over twentieth-century science (1989); yet existing research pertains mostly to the hard sciences and philosophy. My study follows their conclusions, stating that, in the twentieth century, probability ceased to function as a mere heuristic instrument and became a concept necessary for understanding the materiality of physical phenomena, communication and cognition. Building upon their work, I assert that traces of a probabilistic world-image can be found also outside the confines of the hard sciences, in the aesthetic laboratories of the avant-garde. Here new subjectivities and sensibilities were being forged to adapt modern man to life in new social, political and economic environments which overloaded the sensory apparatuses with deafening sounds of machines, images of terror, and constant influx of (mostly anxiety-inducing) information from around the globe torn apart by conflicts and socio-economic turmoil. In the conditions of prolonged and irreducible uncertainty, imposing on whole societies the necessity to endure both minor accidents and grand disaster, physical displacements and perceptual shocks, pressures on the mind and the senses to adapt grew exponentially.

The idea that one can perceive and comprehend reality in terms of shifting probabilities emerged at the same time in the work of many avant-garde artists in Western Europe and North America. If scientists in the twentieth century strove to produce knowledge that was not necessarily true, but (only) highly likely, artists followed suit, though in the opposite direction, looking for improbable forms, behaviours and associations of ideas. In so doing, both professions redefined knowledge, revealing its fundamental indeterminacy, relativity, future-orientedness and performativity. My aim is to show that, since the beginning of the twentieth century, many avant-garde artists have grasped the growing complexity and acceleration of technological and social change and reached conclusions parallel to their peers in professional laboratories. Artists and writers like Umberto Boccioni, John Cage or Iannis Xenakis came to

Francis Bacon? He rightly noticed an uncanny resemblance between Bacon's paintings of warped, dissolving, almost indiscernible faces and the distinctive aesthetic of images generated by GANs, which probabilistically create new images after being trained to discern existing patterns in databases collecting human-made art. It is, however, equally fascinating to ask the following: How is it possible that Bacon's twisted imagery looks like the probabilistic art of GANs, programmed long after his death?

understand that randomness and uncertainty are part of the very fabric of the material world, while the seemingly perfunctory act of prediction – calculating and adjusting probabilities in response to ongoing changes in the environment – constitutes the essence of perception and cognition. Boccioni's obsession with cognitive and perceptual uncertainty, Bruno Corra and Emilio Settimelli's equation of art and economic scarcity, the surrealists' practices of randomisation for exploring the unconscious, Marcel Duchamp's relational and probabilistic theory of art, John Cage's concept of indeterminate performance, or Iannis Xenakis's stochastic music – these are the most conspicuous examples of the probabilistic mindset in avant-garde movements. By consciously renouncing common sense, bidding farewell to the past and adopting the role of secular soothsayers, these artists were able to forge probabilistic forms of seeing and being in the world.

The examples chosen for this study thus extend across five long decades of great turmoil and almost unbound, or to use a more apt word, 'unhinged' creativity from the beginning of the First World War to the decade following the end of the Second World War: from the Italian Futurists, intoxicated by the violent symphonies of artillery, to the singular case of Iannis Xenakis, a victim of war whose great ambition became the task of translating the auditory chaos of modern warfare into classical composition. These artists stemmed from different cultural backgrounds, expressed different attitudes toward modernity, displayed various political inclinations, ranging from fervent and aggressive nationalism, through detached, post-national liberalism, to enthusiastic techno-utopian anarchism, and – for those reasons – have rarely been discussed together. However, as I want to argue, they can be discussed jointly, not only because of their shared interest in 'taming chance', but also based on their rejection of humanist culture. Their utopian attempts at forging new sensibilities hinged on the negative experience of losing ground – a sense of continuity, belonging, and connection to the past – which alienated them from their respective societies, but, at the same time, created conditions of possibility to speculate about completely new visual forms, auditory experiences, artistic practices and subjectivities on which the former relied. The dissociation of the avant-garde from tradition, and, above all, of the avant-garde artist from customary cognitive and perceptual habits permitted 'unhinging of the senses', to use Beck and Bishop's notion, and their reconnection according to new principles, developed on the basis of various ideological presuppositions, freshly contrived needs of the industrialised society, and experimentation, either directed at precise aims or liberated from any practical goal-orientedness. The conditions of living during the first half of the twentieth century certainly put a lot of strain on societies, unaccustomed, that is, lacking resilience, habits and self-awareness, to dealing with the complexity and speed of the industrial environment. Already in 1903, over a decade before the shock of the 'Great War', Georg Simmel observed that psychic self-estrangement became a prerequisite to inhabiting the urban jungle, so over-stimulating as to enforce on the individual a necessity to develop a cold, rational shell of detachment in which to retreat (see 1971: 324–39).

The sensory overload brought about by technological progress was experienced at its most extreme by soldiers in the trenches of the First World War, who were exposed to relentless artillery and machine-gun fire – a threat to life and pressure on the senses incomparable with anything encountered by humankind in the past due to the power, suddenness and inherent randomness of explosions of shells which travelled at almost imperceptible speeds and from unknown origins.[3] The pressures of mechanised, industrialised violence resulted in mass instances of nervous breakdowns of a special, previously unobserved kind – sensory dissociation – first described as 'shell shock' by Charles S. Myers only six months after the war broke out. The cases collected by Myers displayed, besides hysteria-like symptoms (very troubling for the military personnel for whom such odd behaviours were exclusively female), hallucinations, amnesias, troubles discerning dreams from memories, and – most importantly here – various forms of sensory dysfunctions, like blurred vision, decreased colour sensitivity, oversensitivity to noises, reduced acuity of taste (Myers 1915).[4] All of these symptoms could not be explained by physical injuries and could be reversed through hypnosis or returned to normal after some rest away from the front line. For the avant-garde artists, who flocked to and created almost exclusively in

---

[3] Erich Maria Remarque expressed this state of confusion in his famous *All Quiet on the Western Front* in the following words: 'We lie under the network of arching shells and live in a suspense of uncertainty. Over us Chance hovers. If a shot comes, we can duck, that is all; we neither know nor can determine where it will fall.' I elaborate in more detail on the topic of war and dissociation in the context of Italian Futurism in Wojnowski 2019.

[4] Myers points to the relation between shell shock and hysteria in the following passage: 'It is therefore difficult to understand why hearing should be (practically) unaffected, and the dissociated "complex" be confined to the senses of sight, smell, and taste (and to memory). The close relation of these cases to those of "hysteria" appears fairly certain' (14). It is worth noting that symptoms previously labelled under the term 'hysteria' now belong to the conditions termed in American Psychiatric Association's DSM-5 as dissociative disorders.

great metropolises of Western Europe and North America, and who often experienced the traumatising effects of modern warfare first-hand (as victims or witnesses), alienation from oneself and dissociation of the senses became profound affective states which grounded existing nihilist convictions in the body and sparked interest in self-experimentation in search of new ways of sensing the world.

As perplexing as it may sound, for many artists who fought in or behind the trenches the war often remained a profoundly ambivalent event that sharpened their awareness of beauty, joy and other human emotions as it also opened them to new perceptions of reality (for an overview of such positive accounts of war see Becker 2000). The overheated and disorganised sensorium could be pieced back together, the 'un-hinged', raptured mind reinvented to create and inhibit a new reality. Striking new images sutured from many points of view, materials and surfaces; poems in which a unitary voice gives way to utterances without precise origin, belonging to nobody; even words dissolving into syllables and letters of different fonts and sizes – all those cases of representation breaking down create conditions of possibility for new forms of order, aesthetics and sensibilities to emerge. It is at this point in art history that probability, and other revolutionary concepts from the field of science, make their great entrance onto the scene and penetrate the imagination of artists (and the language of manifestos), allowing them to speculate about the future of human senses. For a lack of an existing term describing this aspect of many artistic movements born in the turbulent first half of the twentieth century, I propose here to group them under the label 'dissociative' avant-garde.

The notion of writing about conceptual links between probability, which rose to its epistemic prominence as a mathematical concept, and hardly quantifiable matters of artistic production demands some explanation. Probability does not describe a property of an object; it refers to the behaviour of dynamic systems of multiple elements. When we study such systems and observe their recurring features we can, using probabilistic methods, predict their future states with varying degrees of confidence. Calculating probability is an abstract, inferential process that involves partitioning the future and imagining events in terms of their possibility and our expectations; it answers questions about the likelihood of events, producing knowledge in the form of the numbers 0 and 1, where 0 indicates impossibility and 1 indicates certainty. If so, how can this kind of quantitative knowledge relate in any way to artistic production, or any other non-scientific sphere of culture?

Well, firstly, despite what has just been said, probabilistic thought cannot be reduced to arithmetic. Probability theory, as it developed from the mid-seventeenth century throughout modernity, was influenced and altered by those areas of science in which it has been applied. Moreover, because attempts at adopting probability in new branches of science were almost always deemed controversial, it was followed by tempestuous debates about the nature of human knowledge, causality, freedom, truth, etc. As the concept infiltrated new areas of science, it was enriched by new formalism, or given new meanings:

> Even when probability has entered at the level of methods rather than of theories, the ultimate impact has transcended technique. When psychologists adopted inference statistics as a tool of the trade, they also came to view the same techniques as models of the mind. When biometricians warred with Mendelians over the proper approach to genetics, the view of biological inheritance implied by Pearsonian statistics was at issue. When physicians opposed the use of randomized clinical trials, they doubted not only the relevance of the results but also the ethics of the therapy. Whatever they touch, probability and statistics transform, and are themselves often transformed in the process. (Gigerenzer et al.: xv)

The fruitful, yet controversial alliance between probability and quantum physics proved to be the most transformative not only for both areas of knowledge, but for the philosophy of science as such. Not only did it break the long rule of determinism, but also it cast doubt on the dominant view that science objectively describes reality. Embracing probability in physics led to conceptualising science as inherently predictive and understanding of most physical processes as inherently uncertain. These epistemic breakthroughs sent shockwaves all through culture, revealing science itself, as an ever-evolving discourse, to be prone to accidents.

In my view, it was exactly these transformations involving probability that made science at the turn of the nineteenth and twentieth centuries so inspiring for artists. When science no longer had to abolish chance (through its saviour, strict causality and determinism) but could 'tame' it with probability, it left space for creative freedom and unpredictable results, even within the confines of the most rigid and intricate mathematical formalisms and the strictest experiments. The probabilistic turn in science coincides with equally, if not more, severe seismic events in Western culture, during which art freed itself from its long-running obligations to the familiar media, aesthetic values and mimesis. Some artists provoked and amplified these seismic shocks by exhibiting black squares instead of the familiar landscapes; others by insulting and provoking the public instead of offering something decent and well-contained ('in-itself') to look at or listen to. The avant-garde developed and mutated in many, often opposing directions – formally, aesthetically, ideologically, etc. – but in all the artistic circles of the time, among those who took their duties with the utmost seriousness, as well as those who tackled them light-heartedly, arose a shared recognition that without a common approach and a simple sense of purpose (to depict!) art reveals itself to be a kind of childish game for those who feel disillusioned with grave ideals and sombre forms, and a great war for those who handle brushes and easels as if they were machine guns and tanks (see Getsy 2011). The playfulness of the Dadaists or strategic cunning of the Futurists both adapted to a common experience of groundlessness: a lack of determinate rules on how, why and for whom to make art dictated by the authorities of the past, coupled with a sense of immense pressure to perform, either economically (to earn a living) or symbolically (standing out in the anonymous crowds of mass society). Irrespective of their goals, the artist had to learn how to surprise and draw publicity, how to predict the expectations and strategies of others (artists, critics, the public), and how to experiment with and master chance – to astonish even themselves (and be ahead of the game).[5] Moreover, purposefully going against the rules and expectations of the old masters and the general public turned the spotlight on these very same rules and expectations and held them up for debate, raising artists' awareness that art can be understood and practised as a game.

In other words, being an avant-garde artist demanded more than craftsmanship, that is, knowledge of how to fit one's imagination into frames and moulds, literally and metaphorically as the fixed expectations of the aristocratic patronage. When defying expectations became a goal in itself, the whole rationale of being an artist had to change accordingly. Instead of simply fixating on making beautiful (or truthful, or morally valuable) things, one had to learn to think and act strategically, to master the logic of 'operations'. One could push the boundaries of art and venture to new and uncertain areas:

> An operation, rather than a work, does not leave untouched the sphere that attempts to regulate it. The evident works (the writings and exhibited objects) no longer settle easily under the association of art with questions concerning beauty, or fall prey to the confusion between the artwork and the object or the feeling it supposedly represents or expresses. (Bishop and Phillips: 5)

---

[5] For a fascinating account of the avant-garde's dirty tricks and fraudulent marketing – from price-fixing to violent publicity stunts – see de Chirico 2001.

Without any deterministic laws (rules) of excellence in place, to achieve victory in the messy play-/battleground of art one had to abandon hard truths and learn to think in terms of probabilities. This is because, as noted back in 1832 by Carl von Clausewitz, the great military theorist who was, incidentally, one of the first great intellectuals to notice the importance of probability beyond a relatively small circle of scientists, war creates conditions of great uncertainty and requires one to take advantage of this fact:

> It is now quite clear how greatly the objective nature of war makes it a matter of assessing probabilities. Only one more element is needed to make war a gamble-chance: the very last thing that war lacks. No other human activity is so continuously or universally bound up with chance. And through the element of chance, guesswork and luck come to play a great part in war. (2008: 85)

Unlike the scientist who masters probability to find order in what initially appears random, the general and the artist face more elusive challenges in their respective professions which render precise calculations practically impossible. Contrary to statisticians who search for larger patterns to form expectations, artists (and strategists) scavenge around the margins of probability distributions to find the most improbable strategic operations (and images, sounds, associations . . .), or, in very rare cases, to break past the limits of what seems possible as an art form or war strategy. As I will show in this book, bits and pieces of the probabilistic discourse in science, borrowed (or marvellously intuited) played an important role both in forging new ways of seeing and listening to the world, and in this multifaceted process of redefining art as a game which involves assessing probabilities.

Before I proceed to art-related matters, I want to offer a short introduction to the history of probability calculus as it emerged in modernity, spread slowly and steadily through the eighteenth and nineteenth centuries, and finally ascended to the epistemic throne at the beginning of the twentieth century. As there are great studies of the topic already available, I will focus solely on a few particular moments in a longer history, ones I found of particular importance in terms of culture or crucial epistemological concerns; when probabilistic calculus allowed for new practical solutions to problems and in that way influenced society, or, when it stimulated philosophical discussions concerning key existential problems which become approachable from a completely new probabilistic perspective. This overview should allow the reader to better understand how probability theory – a set of mathematical axioms – can be later related to aesthetics. Some of the problems sketched here will be covered in more detail in the following chapters. To sum up these historical considerations, I proceed to an overview – narrated from the perspective of two great futurologists, Vilém Flusser and Stanisław Lem – of the special role of the concept of probability in modern science and some further speculations on how these seemingly mathematical advancements translate to other problems of anthropological significance.

Probability: From Rolling Dice to Spinning Atoms

There are many reasons why probability calculus – and the modern concept of probability – was not invented until the second half of the seventeenth century, even though games of chance, like dicing, have been present in many civilisations for so long that some scholars hypothesise it could have even been human societies' very first invention (David 1962). If the capacity to calculate odds gives a player a considerable advantage over their opponent, this makes it so much more striking that calculated probability has been absent from human history for so long. Some causes of this state of affairs are related to the inner dynamics of mathematical progress (the lack of the requisite formalisms, such as the limit theorem necessary to undertake such a task), some of them are socio-economic, and some, in one way or another, are connected to the long rule of Christianity, which not only imposed onto European states the doctrine of determinism (the divine omniscience of one God), but also declared games of chance to be an unholy activity, whether they were played or merely theorised about.[6] Hence, many stars had to align for probability theory to emerge in its modern form. Ian Hacking, the pope of probability history, enumerates the following five factors as the most crucial:

- People of power and influence attended to the statistics of births and deaths derived from data that had long been available, but never used.
- The mathematics of gaming appeared as a topic in its own right.
- There arose a new model for assessing evidence in legal disputes.

---

[6] In the epic *Mahábarata*, there are traces of probability calculus being used in ancient India. The third book tells the story of Kali, a demi-god of dicing, haunting the righteous Nala, the king of Nishadha Kingdom, who chose a mortal to become his wife. To avenge this insult he arranges a game of dice, during which Nala gambles away all his wealth and kingdom to his brother, Pushkara. Expelled from his home, he encounters a sage proficient in the 'science of dice' (translation of H. H. Milman [1860, p. 76]) who helps him outsmart Kali and win back his demesne.

- The reliability of testimony was calculated, the possibility of miracles having happened in the past, as reported, was measured.
- There were new proofs of divine benevolence. Bizarre to our eyes – except that the authors showed that they well understood how to test statistical hypotheses from the word go, a conception that had never existed in human thought before. (2006)

Yet, despite many obstacles, it remains surprising that quantitative probability was not invented for so long. Not only had people played with dice and generated stable frequencies in the process, they had also made inferences about the future long before any of these inventions. However, they did not study either of them under the same umbrella, nor refer to them as probabilities. It is very important to stress that, even though the word 'probable' had been widely used before and the concept of probability was the subject of a great deal of philosophical speculation, from Antiquity to the Renaissance, it received no numerical interpretation.[7] Throughout the ages, the term was mostly applied in evaluating opinions in terms of their plausibility, rationality or believability, but with no consensus on how to assess them (on ancient and pre-modern views on probability see Byrne 1968; Wohl 2014). For scholastic philosophers, who grounded their observations on classical philosophy (mostly Cicero, Boethius and later Aristotle), the notion of probability was associated with secondary knowledge, that is, statements that could be reasonably contested with counterarguments. Probable knowledge was inherently imperfect and unsuited to logical formalisations that would ensure their truthfulness, yet worthy of consideration by educated people. With the exception of simple, sensory observations ('the grass is green'), statements by the uneducated were excluded from the domain of the probable. Scholastic philosophers basically agreed that wisdom of the writer was crucial in judging such assertions, though this was understood specifically as expertise in the field to which a given opinion belonged: 'The category of the "wisest" in medieval definitions of probability was a stand-in for "expert", including top-ranking lawyers, theologians, medical doctors and philosophers – or even architects, engineers and other *virtuosi* of mechanical arts' (Schuessler 2019). In other words, one's epistemic credibility relied heavily on 'hierarchical success in academia'. This view was radicalised by the Jesuits in the sixteenth century, who established the casuist doctrine of 'probabilism' (see Tutino 2018). This stated that, in cases of moral uncertainty, one might follow a probable opinion

---

[7] The term 'probabilism' refers to the lack of absolute certainty.

backed by the authority of the Church, even if more probable opinions existed.

Whoever we choose as the originator of modern probability, or whatever we consider its foundational event, consensus remains that it began somewhere around 1650 and many scholars point toward Blaise Pascal as the central character in the plot. Pascal (1623–62) was an avid critic of probabilism and the Jesuits – the pope's most loyal and fundamentalist soldiers warring with the Reformation – who were fully aware of the epistemic and ethical challenges to the rule of the Church over people's hearts and minds. The seventeenth century marks the beginning of the slow decline of the Catholic Church as an undisputed moral authority and a considerable political power. It is at this moment in history that atheism enters the public sphere, especially in England and France, and Christian thinkers find themselves in a completely new position: fending off dissidents. Pascal's famous 'wager', a philosophical dialogue between a fervent believer and a sceptical atheist, which involved an important piece of logical reasoning in probabilistic terms, was motivated by this unprecedented cultural shift. His ingenuity was all the more impressive in that probabilistic arithmetic had not yet been established. In fact, Pascal, along with his fellow mathematician, Pierre de Fermat, is often accredited with inventing calculus, though, as Hacking has stressed, their invention was rather modest compared to the work of Christiaan Huygens, who picked up the topic after visiting Paris in 1655 and only two years later published the first major mathematical work on games of chance – *De Ratiociniis in Ludo Aleae* (On Reasoning in Games of Chance).[8] Pascal never really refined his arithmetic explorations into probability, which were limited to the problem of division of stakes in an interrupted game, but he became instantly aware that probabilistic reasoning could be applied to seemingly incalculable matters of eternal life. Significantly, the concept of probability found its way into existential considerations before it was successfully applied in any scientific or practical endeavour.

Contrary to popular perception, God's existence is not at stake in Pascal's wager (2006). Weighing odds is a rational procedure and, as the philosopher begins by arguing, God must remain entirely inconceivable for the human mind, for it is finite, while He is not. This does not imply, however, that believing in an irrational God is hopelessly

---

[8] The pioneer work by Gerolamo Cardano who wrote 'Liber de ludo aleae' (Book on Games of Chance) around 1564 fell into obscurity for almost a century and was not published until 1663 (Kendall 1956).

irrational in itself. Pascal argues that even if, in principle, we cannot know anything about the Creator's true nature, we can still rationally decide whether to believe in Him or not. What's more: we *must* wager if God exists or not. 'It is not optional. You are embarked', he writes, and the choice facing us is binary, like picking heads or tails in the simplest gambling game. Given God's irrational nature, both answers are equiprobable. For the sake of argument, devised for a critical mind that will not fall for *quasi*-empirical accounts of miracles (these could be disproven), he assumes it is true. He thus diverts the rational attention to reward and expectation, and it is here Pascal finds belief to be a firmly rational choice. He reasons that if one decides to believe in and worship God, one will be given an infinite reward: the eternal happiness of the afterlife. In turn, if one decides not to believe, one risks eternal damnation for small (finite) gains. Therefore, even if one does not believe, if one is rational one should choose to act as if God existed – pray, attend Masses and read the Bible – not because it is certainly true, but because it is more profitable. In terms of Pascal's probabilistic rationality, belief becomes a matter of mercantile calculation – it is the merchant's existentialism. True piousness will eventually come through mechanical repetition. In Pascal's case the central concept of the wager connects three remote areas of knowledge: betting, economic rationality and religion. Such correlation would be difficult to establish in the past without noticing the potential of uncertainty in a rational, calculated manner.

It is no coincidence that early developments in probability caught also the attention of forward-thinking policy makers in Holland, a country that was undergoing rapid modernisation under republicans who had seized power from monarchists and expanded their global influence. Johan de Witt, the Grand Pensionary of Holland (a post combining the offices of Prime Minister and Foreign Secretary) and Johannes Hudde, the Mayor of Amsterdam, both proficient mathematicians themselves, read Huygens's first major treatise on probability and pondered its application to pension planning. They even consulted the author; their exchange broadened the scope of probability's practical applications beyond gambling. Huygens himself became aware of this opportunity as he was presented with John Graunt's *Natural and Political Observations Made upon the Bills of Mortality* (1662) while writing his own book. He praised Graunt's research and saw it as somehow connected to his own study, though he hesitated to comment at length on the possible applications of his theory of expectation in data sets other than gambling results. Nevertheless, in 1671, he was called upon to consult inferences de Witt and Hudde had drawn from Graunt's book. The politicians were interested in probabilistic analyses of life expectation and their potential use in selling annuities – a type of contract between a buyer who wants to secure their future by annual payments of a smaller sum, and a seller, a city council or other public institution, for instance, which instantaneously receives a larger amount of money in return. Such contracts were concluded both with the public good in mind, as a form of pension providing citizens with living support and stability, and as a form of repairing public finances. De Witt published *Waerdye van Lijf-renten naer proportie van Los-renten* (The Worth of Life Annuities Compared to Redemption Bonds) in 1671 as a semi-political, semi-mathematical treatise that supported the political case for other state officials to back his plan to raise money for Holland's war efforts against the British.

De Witt's idea of approaching matters of public policy in terms of gaming drew on Graunt's analyses of mortality charts, providing him with objective mortality rates, and Huygens's theory of expectation, which defines parties of a contract as players betting either for or against one's survival and entering the contract with different expectations. This problem was the first to expose the duality inscribed in the modern notion of probability – its simultaneous reliance on conceptually irreconcilable notions of (objectivist/ontological) frequency and (subjectivist/epistemological) belief. By the first approach, probability describes the relative frequency of an event within its possibility space, for example of a die coming up '1', which has nothing to do with our conscious expectations. If the die is perfectly weighted, it can be seen that the number '1' occurs, on average, once every six throws. The subjectivist perspective, in turn, considers probability theory as the measure of odds that are favourable in relation to those that are possible. In other words, it is the degree of belief that an event will take place. From this point of view, probability does not pertain to anything but our knowledge; for example, one's confidence that a '5' will appear with the next throw of the die.

De Witt's work, despite its limited reception among his contemporaries, marks an important threshold in the history of insurance markets and emerging global capitalism: his probabilistic approach was an important step towards the quantification and financialisation of uncertainty (see Cannon and Tonks 2008).[9] Helga Nowotny provides an apt description of this process which later evolved to

---

[9] The end effect of this process can be observed now in the form of what Ulrich Beck (1992) and Anthony Giddens (1990) call the emergence of 'risk society': a paradoxical socio-economic paradigm in which prediction and securing the future (making the world we live in more predictable) through accelerated techno-scientific progress become key for socio-economic growth; yet the very

almost universal reliance on mathematical tools in the calculation of risk and financial speculation:

> Historically, it took a long time for risk to emerge as an unknown that could knowingly be dealt with. It had to be lifted out of an obscure and potentially threatening mire of numerous dangers that rendered life insecure and posed imminent and real threats. It came with the transformation of the unknown into something known by converting danger into a risk that could be calculated and hence contained . . .
>
> A few centuries later, and based on the same idea of converting a danger into a risk that could be calculated and supported by vastly improved statistics and probability theory, insurance as a regular business took off. Continuously updated calculations of the probabilities of various kinds of risk allowed for adjusting the risk premium. (Nowotny 2015)

It was not only the markets that benefited from the new mathematical tools.[10] During the following decades, at the turn of the seventeenth and eighteenth centuries, as probability theory was further refined, statistics began to slowly emerge as an academic discipline (at first with no relation to mathematics) and collecting data about citizens contributed to the evolution of modern governance. Hacking even argues that the very form of the modern state would be inconceivable without some form of statistical study of the population – from the simplest censuses for taxation and army recruitment, to the more complex practices of contemporary democracies which rely on polls and surveys (to a fault):

> Every state, happy or unhappy, was statistical in its own way. The Italian cities, inventors of the modern conception of the state, made elaborate statistical inquiries and reports well before anyone else in Europe. Sweden organized its pastors to accumulate the world's best

---

same progress leads to new environmental hazards which, in turn, make those predictions more volatile. In addition, because of contemporary practices of stock market speculation and the emergence of various financial instruments, the economic system becomes increasingly dependent on capitalising on various forms of uncertainty. None of these socio-economic changes could be possible without refined mathematical instruments for assessing risk (for an overview of probability and economics see Hamouda and Rowley 1996).

[10] Whether their widespread and widely applauded application for statistical analyses in banks or for predicting shifts in economic trends surmounted to any economic stability is another question. In recent history, financial tools constructed on the basis of probability models in order to manipulate risk have presided over catastrophic failures due to their performative power to further the complexities and volatilities of markets which they were employed to predict for critically inclined and philosophical investigations into the performative affordances of probabilistic mathematics in the contemporary economy and society (see Mackay 2014).

data on births and deaths. France, nation of physiocrats and probabilists, created a bureaucracy during the Napoleonic era which at the
top was dedicated to innovative statistical investigations, but which
in the provinces more often perpetuated pre-revolutionary structures
and classifications. (2006)

Until the eighteenth century, however, attempts at counting citizens were scarce, and incidental to the proper functioning of the state.
The first country to take full advantage of censuses and probabilistic estimations of population was the newly founded Prussia, which
incorporated Brandenburg and which, in a matter of a mere two
centuries, ascended from a minor country dependent on the Polish
Crown to a global superpower. Interestingly, bureaucratic innovations and statistically oriented (quantitative) governance played a role
in its success story. It has been even argued that probability theory
– effectively used by Gottfried Wilhelm Leibniz to estimate its population – contributed, to some minor extent, to the very foundation of
the state. In a memorandum written in 1700, Leibniz advocated for
Prince Frederick to become the king of the united monarchy, stating,
on the grounds of his probabilistic population estimates, and contrary to popular opinion, that Prussia was as powerful as Brandenburg (Leibniz 1864–8: 5, 303–15). As such, its ruler could and ought
to be equally considered for the throne in the election process. In the
decades to come, the Prussian state only embraced population studies
further, establishing a statistics office and attempting a comprehensive
categorisation of its citizens into distinct classes (in relation to land
ownership) for efficient tax plans. At the same time, the University
in Göttingen introduced its first courses in statistics, and Gottfried
Achenwall gave the discipline its name: 'Statistik', meaning knowledge about the state (*Staat*) (see Hacking 1990: 24–6). It is important
to notice that every case of social data collection is related also to
predicting citizens' future behaviour and consequently undertaking
some form of organised, top-down influence, either by conserving,
encouraging or counteracting certain patterns.

This new branch of knowledge came in even more handy during
the rule of Frederick the Great, famed for his relentless passion for
bureaucracy and governing his country like a well-oiled machine. To
counteract the individual agency of his officials, still embedded in the
feudal system and granting nobles too much freedom in supervising
their provinces, Frederick introduced collective boards of clerks who
were tasked with preparing detailed reports on the state of affairs
in their territories and presenting them before their ruler. This collective form of governance was not motivated, by any means, by
Frederick's democratic inclinations. On the contrary, the sole reason

for their introduction was, as noted by Walter L. Dorn, 'fettering the individual civil servant to a collective board, (. . .) by subjecting both to bureaucratic regulations of imperative and canonical authority' (Dorn 61). Precise calculations of data and removing the individual bias in the decision-making process were essential for:

> this complicated meshwork of boards and individual officials [that] had the supreme function of collecting data, correlating facts, drawing up balances of trade or contracts for leasing the royal domains, preparing budgets and drafting statistical reports, in other words of supplying the autocratic king, the highest civil servant and only real man of action in the realm, with the indispensable information on which to base his decisions. (62)

Frederick's innovations were a prime example of a more widespread trend in European politics in the eighteenth century – to expand disciplinary measures and governmental control over the population. Michel Foucault described this process in a series of lectures he gave at the Collège de France in 1978. He took notice of the role probability calculus might have played in forming what he calls 'biopower' – a new administrative rationality which expands political power over new domains of human life. One of Foucault's prime examples is the early practice of inoculation and variolisation against smallpox – primitive methods of immunising people by exposing them to small amounts of infected material taken from a sick person, first introduced in Western Europe in the mid-eighteenth century, but without any theoretical understanding.[11] However, as Foucault cleverly points out, this lack of scientific (i.e. causal) foundations prompted some intellectuals to 'think of the phenomena in terms of the calculus of probabilities. To that extent', he continues, 'we can say that variolization and vaccination benefited from a mathematical support that was at the same time a sort of agent of their integration within the currently acceptable and accepted fields of rationality' (2007: 58–9).

We might find it surprising that, long before Louis Pasteur's germ theory of diseases, there was an understanding of epidemics in statistical terms which has even withstood the test of time. One avid supporter of inoculation Foucault mentioned by name was Daniel Bernoulli, whose work was recently acknowledged as 'ahead of modern epidemiology'. In proposing to analyse mortality rates among smallpox victims with probability calculus, Bernoulli was the 'first

---

[11] Inoculation to prevent smallpox was practised earlier in China, Turkey and India.

to express the proportion of susceptible individuals of an endemic infection in terms of the force of infection and life expectancy' (Dietz and Heesterbeek 2000: 513). By the same token, he was one of the first thinkers to reflect on social problems in terms of probabilistically expressible regularities and constants.[12]

Bernoulli based his method on the work of his relative, Jacob, who had introduced and formalised many key concepts in probability theory eighty years earlier (he wrote his seminal text on the topic, *Ars Conjectandi*, between 1684 and 1689, though it was published posthumously in 1713). As the title implies, it talks about the 'art of measuring the probability of things as exactly as possible, to be able always to choose what will be found the best, the more satisfactory, serene and reasonable for our judgements and actions' (2005: 10). Bernoulli's general intention was to show how probabilistic reasoning could be applied to various real-world problems in which a rigorous method is necessary, despite insufficient information: by scientists, moralists and lawyers alike. However, his most important contribution to the history of mathematics lay not in presenting a comprehensive philosophy of probability, but in defining the first limit theorem of probability calculus, commonly known as the law of large numbers. The law expands on Huygens's theory of expected value and describes the relation between the number of experiments and the factual probability of a random event happening: more tosses of a fair coin leads to a greater likelihood that the probability of heads turning up actually equals 0.5. For example, if one throws a coin only ten times, it is not entirely inconceivable that the result will come out the same, let's say 'tails', ten times in a row.[13] However, if we throw the coin a million times, the likelihood of always getting the same result becomes infinitely small. Bernoulli also used an example which would prove particularly inspiring for

---

[12] Interestingly, Bernoulli's discoveries prompted a heated dialogue between him and Jean le Rond d'Alembert, who argued that probabilistic calculations are of little value in general because people tend to bet on large long-term risk instead of small short-term. In other words, d'Alembert was rightly convinced that most people try to avoid immediate danger instead of looking at the bigger picture (see Daston 1979). His point seems to hold pretty well if we take into account recent mistrust toward vaccination against coronavirus SARS-CoV-2, an incomparably safer medical procedure in comparison to early forms of inoculation.

[13] In 1913, in Monte Carlo, such an event actually took place: the roulette table came up black twenty-six times in a row. This led many players to frantically bet on red after the fifteenth consecutive 'red', because they wrongly believed that the probability of the next result was tied to the previous series. This was obviously untrue, and led many players to lose fortunes at the table.

certain physicists a hundred years later. An urn is filled with pebbles of different colours, mixed in initially unknown proportions; someone draws one pebble at a time from the urn, records the result, and puts it back so that the picks are independent and do not affect the pebble ratio. The theorem states that as the number of experiments nears infinity, the probability that the observed proportion of balls of different colours will agree with the true proportion moves toward certainty (16).

Bernoulli's work turned out to offer solid foundations for subsequent expansions of probability calculus in many different areas of science, because 'it linked the probabilities of degrees of certainty to the probabilities of frequencies, and because it created a model of causation that was essentially devoid of causes' (Gigerenzer et al.: 29). To put it differently, Bernoulli used combinatorics to create a mathematically (and epistemologically) sound model that connected observations of frequency with mathematical expectations, allowing probability theory soon to conquer new branches of science. However, he admitted modestly, in truth, that he only provided a precise mathematical formula for a common-sensical observation, '[b]ecause, even the most stupid person, all by himself and without any preliminary instruction, being guided by some natural instinct (. . .) feels sure that the more such observations are taken into account, the less is the danger of straying from the goal' (19). Yet he inadvertently invented something of grand significance: a method for dealing with the regularities found in nature without any knowledge of the underlying microscopic processes. Equipped with Bernoulli's formulae, it became possible to evaluate certain data sets, such as mortality or infection rates, and form predictions without a deeper understanding of the phenomenon (for instance, why people are prone to sickness at certain ages, or what the true cause behind epidemic outbreaks is). With few or no tools to investigate complex problems in-depth, scientists could still try to grasp them merely by looking at the bigger picture.

The possibility of inferring causation-less conclusions hinged on obtaining sufficiently vast pools of data. Therefore, in the decades following the mid-eighteenth century, probabilistic reasoning was nurtured and developed in producing vast contexts for the physical and social sciences where the necessary data sets – astronomical observations, or criminal records – were already available. The leading innovators connecting probability with these seemingly remote spheres of research were two francophone mathematicians: Pierre-Simon Laplace and Adolphe Quetelet, both of whom started their scientific careers in astronomy. While the former made crucial mathematical advancements in probability calculus, the latter was among the first to successfully implement probability in the social sciences, laying the foundations for sociology. It is largely through their efforts that probability became not only a widely used and discussed concept in the scientific community, but also a subject of keen interest to the general public. And from this point on the pace in which the concept travelled from problem to problem and from discipline to discipline accelerated quickly and began to bring them together, leading at the same time to erosion of some of its core foundations.

Laplace became preoccupied with probability in his search for novel methods of combining observations – beyond simply taking the mean – needed for his studies of the stability of planets in the solar system. He employed calculus to try to resolve the inconsistencies in linear equations describing the motions of Saturn and Jupiter – perplexing errors in observations which led Isaac Newton to assume God's intervention was behind the solar system's apparent overall permanence. In 1787, after years of investigations, Laplace proved there was no need for such a lofty hypothesis and these observable inequalities actually proved to be periodic in the long run. Examining data collected on Saturn's position over a 200-year period, he found a remarkably simple property explaining the ratio of Saturn's retardation and Jupiter's acceleration. On the basis of probabilistic data analysis, he was able to predict the planets' movements without any knowledge of the precise causes determining them. He considered anomalies in the movement of the planets as one would frequencies of dice rolls, or the ratio of boys and girls being born. Referring to Bernoulli's work, he treated the true value in probability distribution to be an unknown cause, and three observables as effects (for the best discussion of Laplace's method, see Gillispie 1997: 14–28). 'Laplace's success in this celebrated problem was in considerable part a statistical triumph, a model of the ways in which analyses of data suggested by theory may in turn suggest hypotheses requiring further theoretical development and then observational confirmation' (Stigler 1990: 31n). His interests in probability and celestial mechanics were thus closely tied, the former applied to fill in the gaps in knowledge we could gain about the latter. However, Laplace never believed in reality of probabilistic laws; that is, he would never agree that the universe could be, to some extent, random. In 1776, the same year he graduated from École Militaire and began his career as a scholar, he stated his view on the role of the probability calculus in the physical sciences:

> [I]t is important to pin down the sense of the words *chance* and *probability*. We look upon a thing as the effect of chance when we see

nothing regular in it, nothing that manifests design, and when furthermore we are ignorant of the causes that brought it about. Thus, chance has no reality in itself. It has nothing but a term for expressing our ignorance of the way in which the various aspects of a phenomenon are interconnected and related to the rest of nature. (1776: 14, in Gillispie 1997: 25)

He repeated the same sentiment in his late work, *A Philosophical Essay on Probabilities* (1814), intended for the educated general public, adding an unlikely and impressive eulogy to imperfect but necessary probabilistic knowledge.[14] Laplace expounded the broad spectrum of uses of probability calculus: from celestial mechanics to practical judiciary problems, evaluating testimonies and estimating the optimal number of judges in a tribunal to secure a fair judgement. Moreover, assuming that 'all knowledge is problematical', he asserts that 'the principle means of ascertaining the truth – induction and analogy – are based on probabilities; so that the entire system of human knowledge is connected with the theory set forth in this essay' (1902: 1n). It is worth recalling that it was Laplace who introduced the powerful metaphor of the all-knowing demon, a hypothetical creature who, having mastered everything about the state of the universe at a given moment, can infer its past and present as they evolve according to immutable laws of Newtonian mechanics. While remaining an ontological determinist, Laplace presented himself as an epistemological probabilist.

Working on celestial dynamics, Laplace invented another important model – double exponential distribution, which was later named after him, sometimes also referred to as the first law of errors (Wilson 1923). This law states that the frequency of an error in an experiment is equal to an exponential function of the square of that error; it can be used to approximate questionable data sets from astronomical observations (assuming that each attempt is subject to errors that are independent but of comparable magnitude). His work also laid the foundations for the second law, or the 'normal' distribution, named after Carl Friedrich Gauss, who gave it its full mathematical expression. As mathematical laws, these distribution models eventually found their way into the moral sciences and

---

[14] Another interesting aspect of Laplace's engagement with probability is his devout work for the state during Napoleon's rule, which earned him the post of Minister of the Interior for a short period of time. Laplace even dedicated some of his work on probability to France's political leader, also his former student, whom he passed, luckily, during an exam at École Militaire in 1784.

proved to be very important resources in establishing 'social physics' as an academic discipline in the nineteenth century. This was possible because it so happened that some data sets available to scientists at the time – mostly concerning the distribution of certain physical traits – fit perfectly into the mathematical model.

One of the most important figures responsible for the mathematisation of the science of mankind was Quetelet; after having founded the Royal Observatory of Belgium in Brussels (and giving up on a career as a poet and playwright), he went on to pursue a career as a social scientist, publishing pioneering analyses of social phenomena like suicide and crime, and establishing networks and associations for the first sociologists. Laying the foundations for a methodologically rigorous and mathematically informed science of society, he conceptualised social bodies, comparing them to the celestial ones, which move 'along some predetermined route(s) but (are) subject to perturbations' (Porter 1981: 83). Porter aptly summarises Quetelet's ideas on this:

> According to Quetelet's astronomical metaphor, the development of chest size, of height, and of propensities to crime, suicide, and marriage were all subject to constant and perturbing causes. The constant ones acting alone would have produced the notorious 'mean man' who played the central role in Quetelet's social physics. Perturbing causes distorted this golden mean, however, generating a distribution that could be likened to the deviation of stellar measurements in astronomy. Just as a telescopic observation was subject to a host of variable sources of error arising from atmospheric distortion, expansions and contractions due to temperature changes, and the imperfect vision of the human observer, so all men deviate from the golden mean. (92)

Quetelet was drawn to topics of moral controversy for a very mundane reason: besides the most basic data on body measurements and mortality, crime statistics were the only form of social data available. At the time, there were no public opinion polls, nor other practices of interviewing citizens. To look at society from a distance, one had to focus on the body, death or wrongdoing, which 'offered quantifiable frequencies as predictable as any found in observational sciences' (Donnelly 2015). In this data, Quetelet found recurring patterns, which he modelled and represented using probabilistic distributions borrowed from Laplace and Gauss. For him, the accordance between data sets in astronomy and state institutions was proof that social laws had a physical reality, materially determining individual lives. What appeared inexplicable on its own – a desperate act of suicide or a horrific murder – when considered en masse suddenly turned out to conform to social laws. But the sheer thought that such horrendous and seemingly individual acts might be predictable soon stirred many controversies in public opinion and marks one of the most important early moments in the history of intersections between probability theory and culture. In fact, it is one of most important markers of the deepening crisis of the humanist tradition that relied strongly on the concept of free will.

Quetelet never identified philosophically with determinist positions. His sole intention was to establish a science of mankind which would find and explain regularities occurring in social reality on a mass scale. For example, one of his main interests was the relations between traits of certain populations and environmental factors like climate or topography. He asked simple, seemingly non-controversial – from a contemporary perspective – questions, such as if severe winters can adversely affect human willingness to reproduce, or if geography explains differences in human weight, and tried to answer them with scientific rigour. But even such a moderate approach met with severe criticism and provoked many intellectuals of his time. Even if his theories never really negated the concept of free will, the very intent to speak of humanity without attention to individual agency rubbed many traditionalists up the wrong way. And indeed, Quetelet saw history, as it was practised at the time, as too absorbed in the individual stories and human antagonisms which, in his words, degraded the discipline 'to a portrait of the deplorable effects of [humanity's] rage' (1829: i).

Despite his intentions, mainly owing to the misrepresentation of his ideas by the widely read, controversial English historian Henry Thomas Buckle,[15] Quetelet became widely associated with a new intellectual position, which Hacking aptly labelled 'statistical fatalism' – the conviction that contingencies on a smaller scale conformed to grand regularities and thus human freedom might turn out to be illusory if one's actions are determined by statistical laws.[16] This

---

[15] Quetelet's ideas reached millions of readers all around Europe through Buckle, whose method of historical investigation formulated in *History of Civilization in England* largely built upon the concept of statistical laws. It was Buckle who landed a direct hit on the notion of free will in the book's lengthy philosophical introduction.

[16] This idea features prominently – as a negative point of reference – throughout the literary works of Fyodor Dostoevsky, whose grand depictions of moral decline in Russian society bear an almost sociological quality. Dostoevsky was quite careful to include social and economic backgrounds to explain motivations, yet he always stood by his conviction of the incomprehensibility of individual actions, which cannot be exhaustively explained by any method, let alone a statistical law (on Dostoevsky's views on social science, see Vladislav Bachinin, 'Sociology and Metaphysics in the Work of Fyodor Dostoyevsky'. *Social Sciences*, Vol. 32 (2001). https://ciaotest.cc.columbia.edu/olj/socsci/socsci_01bav01.html. Accessed 16 February 22.)

pessimistic attitude largely owed its emergence to the processes of social, economic and political modernisation in the nineteenth century, an age of great social turmoil, profound class changes, mass migration to fast developing industrial centres and governments' haphazard attempts to make sense of and control the urban masses. Its most important condition was the introduction of the concept of statistical law – an intellectual artefact that had previously been nowhere to be found and used to represent similar patterns of data distribution in a wide range of data sets (Hacking 1990: 121).[17] The idea of statistical laws also reverberated back to the discourse of physics, reviving atomism in physics in the nineteenth century. One might find it surprising today, but the atomist doctrine was not even seriously considered in the scientific community until the mid-nineteenth century, when Rudolf Clausius formulated his two laws of thermodynamics. The revelatory aspect of Clausius's work was to associate heat with particle motion by assuming random particle movements taking place underneath the observable level. This implied the necessity of a mathematical tool to calculate mass particle events: myriads of – supposedly real – gas atoms swirling endlessly in a tank, like the balls distributed randomly in an urn that had grabbed Bernoulli's attention a century before. This idea was widely contested by many scientists at the time, and only became an accepted scientific fact with the publication of Albert Einstein's paper on Brownian motion in 1905. However, it was not Clausius who proposed relying on probability calculus to solve thermodynamic equations, but James Clerk Maxwell. His 'first paper on the dynamical theory of gases opened an era during which physics has become increasingly dependent on the techniques and concepts of probability and statistics' (Porter 1981: 94). They allow to shift the focus in physics from focusing on relations between discernible, large objects of definite properties to indeterminate, complex systems consisting of an indefinite number of smaller entities. Maxwell became attracted to statistical laws and probability distributions in trying to calculate the distribution of gas particle velocities, which Clausius assumed to move at uniform speeds. Building on his predecessor's work, Maxwell proved that these velocities were, in fact, distributed

---

[17] On the other hand, the same concept proved enticing to the proponents of laissez-faire liberalism, providing them with arguments for the further expansion of the free market. From a liberal point of view, statistical laws were evidence of a spontaneous order emerging from the apparent chaos of social interactions. If so, with little state supervision and interference, individual freedom should give rise to different forms of macroscopic, emergent order – an idea very similar to Adam Smith's concept of the self-regulating free market (see Bittermann 1940).

and best represented by the 'law of errors', which Laplace used in astronomy and Quetelet in the social sciences. Although he did not refer to Quetelet's work in the initial paper, in later, more philosophical publications he admitted to drawing inspiration from social physics and the idea of:

> distributing the whole population into groups, according to age, income-tax, education, religious belief, or criminal convictions. The number of individuals is far too great to allow of their tracing the history of each separately, so that, in order to reduce their labour within human limits, they concentrate their attention on a small number of artificial groups. The varying number of individuals in each group, and not the varying state of each individual, is the primary datum from which they work. (Maxwell 1873)

Maxwell went on to argue that he found this method indispensable in dealing with physical processes involving 'large groups of molecules', as it was practically impossible to know everything about the individual states of all the particles in a system. One can only know macroscopic values: temperature, chemical substances, volume, temperature and pressure. About such groups nothing could be ascertained in complete detail, so there was no point in trying to calculate the trajectories of individual particles. Thus, after introducing the molecular perspective into the study of energy transformations, he stated 'our experiments can never give us anything more than statistical information'. However, if we were to abandon this grey area of 'chance and change', Maxwell assures his readers, we would re-emerge in a place 'where everything is certain and immutable', in the domain of the Laplacian demon. In other words, if we turn our attention to singular bodies – large-scale objects, or elementary particles considered individually – uncertainty would be resolved, and the perfect world of determined trajectories would remain safe and sound.

This last assumption did not stand the test of time particularly well, if we consider that statistical mechanics, established by Clausius, Maxwell, and later refined by Ludwig Boltzmann, served as the basis for theory of quanta, which turned uncertainty from a purely epistemological notion into a semi-ontological quality. The successful implementation of probability theory denigrated the image of the universe as a perfect mechanism, carried over from early modern/Baroque philosophy and theology, supplementing it with another useful metaphor: that of games of chance. It was a crucial transition also on a metaphysical level, because one can imagine a maker of this perfect mechanism, whereas game(s) are always bound by an element of chance, making them unpredictable, or – from another point of view – players in a game are kept in a state of unresolvable uncertainty as they cannot fully predict each other's moves (complete knowledge turns a game into a mechanism rendering the metaphor useless). This epistemic change happened involuntarily and quite by necessity, as the physicist responsible for this step, Max Planck, did everything in his power to avoid incorporating statistical methods into his own work on radiation. I write on Planck's unwanted discoveries in more detail in Chapter 5 (on Iannis Xenakis's compositions) and on the probabilistic aspect of quantum measurements in Chapter 1 (on Italian Futurism), so here I will try to be as concise as possible. This is a crucial moment in the history of probabilistic epistemology, because it is at this moment that probabilistic knowledge was elevated from secondary status – as a mere tool applied out of necessity to compensate for practical complications – to earn a vaunted position at the core of modern science. It was for this reason, out of reluctance to incorporate probabilistic reasoning into the hallowed canon of the hard sciences, that Planck spent many years trying to find a theoretical interpretation for his experiments into black-body radiation that did not challenge the dogmas of classical physics. Eventually, he turned to Maxwell and Boltzmann's work, but only out of sheer 'desperation' (see Klein 1966: 298), to discover that radiation, uniformly considered continuous, also exhibits particle-like qualities (this led to the discovery of photons in 1905). This was the first among many new innovations that shattered dogmas of the *ancien régime* in physics in the first decades of the twentieth century. As Richard Feynman aptly sums it up:

> From about the beginning of the twentieth century experimental physics amassed an impressive array of strange phenomena which demonstrated the inadequacy of classical physics. The attempts to discover a theoretical structure for the new phenomena led at first to a confusion in which it appeared that light, and electrons, sometimes behaved like waves and sometimes like particles. This apparent inconsistency was completely resolved in 1926 and 1927 in the theory called quantum mechanics. The new theory asserts that there are experiments for which the exact outcome is fundamentally unpredictable, and that in these cases one has to be satisfied with computing probabilities of various outcomes. But far more fundamental was the discovery that in nature the laws of combining probabilities were not those of the classical probability theory of Laplace. (2)

The first of those strange phenomena was the discreteness of seemingly continuous radiation, unwillingly discovered by Planck, which opened a Pandora's box of mind-boggling discoveries. One of these was spontaneous quantum emissions of energy occurring when particles change their quantum state from excited to 'ground state'. Although unusual from the point of view of classical physics, it is the simplest form of interaction between an atom and an electromagnetic field. These transitions account for most of the light we see, yet they could have not been explained using classical electromagnetic theory. Their discovery marks such an important event in the history of probabilistic epistemology, for they are singular quantum processes that can solely be explained using probability (they do not involve large groups of particles, as in Maxwell's case). This is linked to the fact that spontaneous emissions are not induced by fixed properties of an atom, but by its interaction with an electromagnetic field. According to Paul Dirac's theory of 1927, it is possible to estimate the probability of emission which is an inherently random event. This statement directly contradicted Maxwell's expectation that microscopic processes underlying macroscopically observable thermodynamic changes must behave deterministically.

For some writers, quantum mechanics owes its 'strangeness' not to any particular discovery that is difficult to comprehend, but to the fact that its theoretical core owes so much to probability, which is impossible to visualise in terms of simple relations between distinct objects.

> Quantum mechanics became a strange kind of theory not with Werner Heisenberg's famous uncertainty principle in 1927, nor when Albert Einstein and two colleagues identified (and Erwin Schrödinger named) entanglement in 1935. It happened in 1926, thanks to a proposal from the German physicist Max Born. Born suggested that the right way to interpret the wavy nature of quantum particles was as waves of probability. The wave equation presented by Schrödinger the previous year, Born said, was basically a piece of mathematical machinery for calculating the chances of observing a particular outcome in an experiment. (Ball 2019)

Born's theory builds on Schrödinger's idea that quantum particles also can be treated as waves – a weird concept proven to work mathematically. It aims to explain how quantum mechanics' abstract mathematical apparatus could relate to experimental conditions. Its central claim is that a particle can be assigned a wave function – a mathematical description of a quantum system's state (position of particles in space, or their momenta) in the form of a probability amplitude. Both approaches – particle-oriented and wave-oriented – yield the same results in predicting particles' future behaviour. However, from the wave-oriented point of view, one quite literally can see nothing – there are no observable 'waves', as it is only after the act of measurement that the particle manifests itself to the observer. Born suggested that the amplitude of the wave function is the probability that one will find the particle in a certain position when the experiment is being conducted. This is not to say that it must happen in the first experiment. Rather, we must imagine making a number of experiments (of the same quantum system and using the same equipment) to yield results distributed within a certain space of freedom. In other words, Schrödinger's theory gives us a calculated guess that, in several experiments (out of a larger set), we will tend to find a particle in a certain place. This is unusual, because, typically, in most physics equations, the variables refer to objective properties: the speed of a rocket, or the mass of a falling apple. In Born's view, the wave-function is quite different, as it only hints at what we might expect to see if we observe under certain conditions. It is a probability that, upon taking a measurement, one will find a particle in a given space or with a certain momentum. With Schrödinger's equation and Born's subsequent interpretation, it is often said that physics departed from purely ontic considerations toward epistemic ones: scientific knowledge was no longer solely concerned with objectively existing relations, it had begun to speak about itself. However, the opposite could also hold true, if one assumes that the quantum reality is only describable by a probabilistic theory. The only indisputable fact is that, at least for now, the distinction between ontic and epistemic is unclear when it comes to the results of quantum investigations.

There is no room here to delve into a detailed discussion of quantum physics and its apparent weirdness, as there already exists a plethora of sources on the topic, ranging from the historically oriented (Mehra and Rechenberg 1982), to the philosophical (Omnès 1999) and the popular (Ball 2018). From a probability-oriented perspective, there are two critical issues worth mentioning. Firstly, new physical discoveries in both statistical and quantum mechanics, as well as new advancements in other fields of mathematics, paved the way for a complete refashioning of calculus itself. Secondly, at the beginning of the twentieth century, probability was first linked to philosophical indeterminism, as quantum mechanics 'made the elementary processes in nature indeterministic, and it made probability an ineradicable part of the description of those processes' (Plato 1994: 2). When Schrödinger's discovery shattered the deterministic doctrine professed by the classical thinkers, it automatically installed randomness at the foundation of reality, at its smallest elements – quanta of energy and elementary particles. In 1932, John von Neumann presented mathematical proof that the probabilistic elements in quantum mechanics could not be replaced by a determinist theory with 'hidden variables', as in the case of statistical mechanics, where statistical regularities were explained as the effects of simpler, deterministic motions (1996). At the time, some received his postulate as the final blow against classical physics.[18] However, it was not only quantum physics that played a part in dismantling determinism. For example, statistical mechanics expanded its scope of interest from ideal gasses to include real physical processes involving true randomness.[19] As science became increasingly occupied with physical phenomena of great complexity, it was necessary to find mathematical tools that were not limited by the early concepts and theorems in probability calculus, such as the notion of equiprobability. Jan von Plato explains this necessary transition in calculus:

> The real world does not possess the absolute symmetries of the classical theory's 'equipossible cases.' This was shown by the collection of statistical data, proving real dice to be unfair, having a boy not being as probable as having a girl, and so on. Classical probability is not sufficient for frequentist applications, so that a conceptual change away from the classical interpretation was required. Neither is it sufficient on a mathematical level, so that also a change away from its finitary scope was required. (6)

In fact, the rapid expansion of probability calculus into new fields of science went hand in hand with its refinements in applied and pure mathematics. Measure theory, in particular, proved an essential branch of mathematics in the axiomatisation of probability and subsequent adoption of the concept in all areas of science. Using the measure-theoretic approach, endorsed by Émile Borel and Henri Lebesgue, one could replace the specific concept of (probability) distribution with a more universal one – the set – giving common ground to probability and other mathematical abstractions, such as volume, mass, velocity or even electrical charge. Furthermore, if classical probability dealt only with discontinuous events – like successive throws of a die – then using sets allowed mathematicians to deal with continuous processes and infinitary events. This inclusion of probability under the wider scope of measure theory consequently led to the full axiomatisation of probability theory by Andrey Kolmogorov, eventually making it a

---

[18] It was disputed in later years by Bell (1966), though no definitive conclusions were reached (Bub 2010).

[19] Beyond the advancements in pure mathematics and statistical mechanics, there was also one important innovation in probability calculus which greatly impacted its usefulness – the concept of chained probabilities, introduced by Andrey Markov in 1906.

separate branch of mathematics in 1933. Since then, it has been easier to apply probability calculus in virtually any possible context.

I end my very limited account of the history of probability at this point, because it is here that the rapid expansion of probabilistic logic and concepts into virtually every discipline of science becomes impossible to present in any coherent form, even with major omissions (for an overview of these migrations see Gigerenzer et al.). Instead, I want to introduce Flusser's take on the significance of the probabilistic revolution for the state of human knowledge in general, including humankind's self-image, which meant it played a part in the emergence of the 'post-historical' paradigm (I also return to Flusser's thought in Chapter 5, on Cage).[20] Flusser was keeping a sharp eye on the transformations of post-classical science and on the importance of probability in this process. His interest in this topic was motivated by a very general question: What did this epistemic shift mean for culture, for our shared values and fundamental concepts, like freedom or creativity? He considered the advent of probability theory and statistics in various fields of science as connected with a shift in our perspective on the nature of reality. As the new probabilistic methods enabled scientists to describe the uncertainty and unpredictability of physical processes, playing with probabilities moved from the domain of irrational guesswork to become a pursuit worthy of a learned mind. This also meant that science which parted ways with Newtonian determinism allowed for freedom (indeterminate occurrences) to be conceptualised and calculated as a physical trait. Flusser's broad perspective on the cultural ramifications of the probabilistic revolution in science made it easier to see the importance of probability beyond its application in all the sciences. I will also add one other important historical and theoretical remark concerning the ties between advancements of probabilistic logic in statistical mechanics and the mathematical notion of information.

Human Existence Shaken Like Dice

In one of many grand narratives of the history of mankind, sketched by Flusser in *Post-History*, we find a concise story of the Western civilisation written from the perspective of freedom (2013: 19–22). On a mere three pages, Flusser managed to outline three fundamental ways

---

[20] The following paragraphs on Flusser have already been published in 'Telematic Freedom and Information Paradox' (*Flusser Studies* 2017).

of structuring (governing) existence, which also dictated the basic ways of understanding freedom. The first model was destiny, which supports the religious, teleological world view. The destiny hypothesis states that human endeavours have meaning only within the wider context of God's master plan. Like every other plan, it too has an aim. It is intentional and purposeful – it projects human psychology onto the divine creature, who supposedly created the universe to achieve its goals (whether or not these be known). In the context of Christianity, Judgement Day is the goal which history is steadily approaching. According to this world view, every human deed is counted and receives a meaning as part of a story with a dramatic ending. Destiny as an operative term used to make sense of our existence derives from books – it is a projection of a literary (narrative) rule onto the real existences of believers. In this finalistic image, the question of freedom is posed in the terms of free will. The question is: 'Can man oppose his destiny with free will, and if so, to what extent can he do this?' (20).

The finalistic world view has been complemented in modern times by an alternative: the causal narrative. Flusser associates this new approach with the invention of modern science and its strictly deterministic understanding of nature, symbolically represented by the Laplacian demon. According to this image 'every event is the effect of specific causes, which are in turn causes of specific effects' (19). This image seems even more repressive than the previous one, because here every human conduct is subordinated to the strict laws of Nature. In the mechanical world of Laplace, freedom can only be defined in terms of subjective fallacy – given the impenetrable complexity of deterministic processes, freedom emerges as a state of subjective uncertainty stemming from insufficient knowledge about the true causes of events in which one is involved. What is personally experienced as freedom of choice, in truth follows objective chain-like processes of cause and effect that started long before the subject came into this world. Meanwhile, God the artist-creator is relegated to the far less noble position of the clockmaker-designer.

Fortunately, this terribly pessimistic view on freedom, typical of 'statistical fatalism' in the nineteenth century, was substantially updated after 1900.[21] For Flusser, the probabilistic and statistical formalisations that came to dominate in every field of science gave birth to the 'programmatic perspective'. In this new world view,

---

[21] In truth, both programs could coexist in practice, as few scientists were disturbed by the philosophical implications of statistical determinism. Some, like Maxwell, put a great deal of effort into settling statistical determinism with free will (see Stanley 2008).

chains of cause and effect 'appear only as statistical probabilities' (2011: 141). This breakthrough, which allowed science to speak about reality in terms of probabilities and information, demanded a philosophical shift in apprehending the world: reality itself had to be rendered as the effect of programs (or games). On many occasions, Flusser emphasised that the new advancements in science should have been followed by a new anthropology to redefine man as an improbable 'product of chance' and draw conclusions from this new state of knowledge (1989: 25). This new programmatic world view consumed not only the concept of (fixed) reality, but also languages and meta-languages. From this moment on, virtually anything could be understood in terms of programs, including both of the previous grand narratives: religion and science.

The definition of a program is as simple and elegant as binary code. From the standpoint of information theory, adopted by Flusser from Claude E. Shannon (more on him below), a program is 'every system in which chance becomes necessity', or a 'game in which every virtuality, even the least probable, will be realized of necessity if the game will be played for a sufficiently long time' (2013: 22). Everything from the Big Bang to MS Word, everything that 'computes', that is, turns possibilities into outcomes, can be described as a program.

Flusser's point of view on this matter was also deeply indebted to *Chance and Necessity*, a book by French biochemist Jacques Monod, who wrote this quasi-philosophical study to situate himself against the prevailing perspective that science (by design) strives to explain reality in causal terms compatible with a teleological framework (1971). In contrast to such a philosophical position, Monod argued that 'teleonomy' (goal-directedness) is not in any sense a primary characteristic of the physical world; it is a secondary trait that is a result (and not the cause) of reproduction among living organisms. 'Purpose', 'project', 'cause' and 'effect' are human notions which do not necessarily apply to physical processes, at least not to all of them. Teleonomy remains a very important and useful concept, it is simply not an overarching one. This important belief supported Monod's other claim that stochastic (random) processes should also be viewed as potentially creative. He even went as far as to suggest that it was chance, not necessity, that propels evolution forward, and may have been behind the origin of life on Earth. From this perspective, then, evolution should be understood as a partly random process in which change results from microscopic and non-deterministic quantum noise, disturbing otherwise invariant DNA reproduction. Monod stood up to the hegemony of (purpose-oriented) natural selection as the only viable explanation of evolution. He contested the prevailing deterministic views by praising the role of random mutations as sparks of creativity.

Monod's idea was compatible with Shannon's classic *The Mathematical Theory of Communication*. In the famous introduction to Shannon's text, Warren Weaver stated outright that the amount of information in a message is proportional to the 'freedom of choice, in selecting the message', because 'to be somewhat more definite, the amount of information is (. . .) measured as a logarithm of the number of available choices' (9). For Shannon and Weaver, information, which was 'measured with reference to the number of possible messages that could be sent in a given time using a given set of symbols' (Mahoney 1990: 549), was simply equal to entropy. Entropy, a term first introduced in thermodynamics in the nineteenth century to describe the dissipation of energy in irreversible processes of heat exchange, very soon acquired a statistical interpretation; it was this reformulation by Maxwell that allowed scientists almost a hundred years later to merge the notions of information and entropy.

As I have mentioned, Maxwell proved that every thermodynamic property of a system can be described not only in terms of heat (temperature), but also in statistical distributions. For instance, after we have measured the temperature of a closed system, we can translate this value into the possible number of states of the particles within. At any given time, every particle has a given speed, direction and position inside the chamber. With the rise of temperature (which corresponds to the speed of the particles), there are more possible arrangements of the particles. In fact, even a very small container has an astronomic number of such states: if a particle can have 200,000,000 states, this must be multiplied by 100,000,000, the number of particles inside, to gain the result. Entropy measures the disorder of these particles. In this sense, however, disorder has nothing to do with the intuitive, static meaning of this word. Rather, it designates the possible number of arrangements of particles. When particles have more potential states and are scattered in a larger space, there are more potential states for them to be in. If we relocated all the particles to one side of the container and reduced their speed, their number of locations and speeds would be lower. This example shows how the notion of entropy is closely tied to the idea of freedom. Entropy rises in 'free' systems, where there are more ways for the particles to 'express' themselves. As such, entropy is also related to uncertainty – the more possible states exist, the less certainty we can have of the state at present. Correspondingly, Shannon and Weaver realised that the information value of a message can be measured in relation to the number of possible messages. When we send a long and complex message, it weighs more bits than a short, simple and repetitive one. As Weaver simply put it: 'The greater this freedom of choice, and hence the greater the information, the greater is the uncertainty that the message actually selected is some particular one. Thus greater freedom of choice, greater uncertainty, greater information go hand in hand' (18/19). It seemed very unlikely that an engineer working on a noise-free communication system would admit that noise itself could be a source of information. This meant that squeaks and crackles could provide more information than a boring platitude from someone's relative, just as 'quantum noise' could disrupt gene replication and serve as an evolutionary factor. It also meant an innovative, experimental poem could be more surprising and thus more valuable (information-wise) than another iteration of a Romantic sonnet.

This recognition leads Flusser to his most important thesis, which is also an important ethical postulate: freedom nowadays can be considered in terms of the chance and surprise which occurs during dialogue with others. Such freedom belongs not to the subjective, but to the intersubjective dimension (to the encounter and not the decision). Myths of individualism and other fairy tales about human nature are not only unconvincing from the standpoint of the new sciences, above all, they are politically dangerous. In the efficient hands of public relations specialists, they turn out to be tools of mass manipulation, not guarantors of human rights. Humankind is only left with criticism of programs and constant doubt, which does not mean falling into pessimism or nihilism, because in this game the stakes are freedom of dialogue and the creation of new information. In fact, for the first time in history, humanity stands on the threshold of understanding its situation without being unnecessarily entangled in theological and teleological explanations. As we can see, Flusser, not unlike Pascal, gave probability an existential meaning, but in a completely different sense: not as a method of individual decision-making, but a notion that encourages us to abandon any attachments to free will and find freedom in the joyful and creative unpredictability of communication.

The Polish science-fiction writer Stanisław Lem[22] expressed a similar, though less radical, sentiment about the triumph of probability in scientific discourse in his theoretical opus magnum, confidently titled

---

[22] Stanisław Lem and Vilém Flusser were born barely one year apart. Both were born in Eastern Europe in the 1920s, both came from middle-class Jewish families, and their lives were marked by the experience of the Second World War. Lem was forced to immigrate to Kraków, Flusser wandered the world, never to return to his native Czech Republic. While Flusser condemned himself to eternal wandering, which he made a leitmotif of his philosophy, Lem, faced with the same threat, made the decision to stay. Lem and Flusser were closely tied, however, by something that distinguished them from many other Central European Jewish intellectuals whose lives

Summa Technologiae. Lem's views are in agreement with those of philosophers who took notice that, in the second half of the twentieth century, the foundations of knowledge and scientific practice were built anew, as modern scientists were increasingly dealing with observable and statistically significant regularities instead of pursuing discoveries of new universal laws. Thus, the development of modern science, in Lem's words, was moving 'from singular to statistical laws, from rigid causality to probability, and, as we only understand it now, from simplicity (which was "artificial" in the sense that nothing in Nature is simple) to complexity' (2013: 30). Therefore, physics – and especially mechanics – is toppled from the summit of the hierarchy of sciences, and its place either remains empty, or is being assumed by the information sciences. Scientists today, he explained, gather evidence, analyse data, design experiments and evaluate their validity using computer-run probabilistic mathematical programs.

Lem did not perceive this change as a crisis because, in his opinion, knowledge as information had never really referred to reality, but was always dialectically related to the future. Only now were we becoming aware of this fact: 'Knowledge means expecting a particular event to occur after some other specific events have occurred. One who does not know anything can expect everything. (. . .) Knowledge is thus a restriction placed on diversity; it is greater the lesser the uncertainty of the person expecting something to happen is' (161). If we assume science is a game, the question of whether scientific knowledge has meaning is posed incorrectly. In *Summa* we find an elegant sentence summarising the essence of the problem: 'You cannot ask about the truthfulness of chess' (173). You can only verify if the game has been played correctly, that is, according to the rules. In an age of computation, knowledge is used for action (prediction) and not for contemplation.

Moreover, the transition from 'causality to probability' in science is not merely a technical problem, strictly of note to scientists, engineers and philosophers of science. The stakes of this epistemological shift were, and remain, much greater. Jerzy Jarzębski writes:

> *Summa Technologiae* (. . .), as evidenced by the title, borrowed in part from the writings of St. Thomas Aquinas, and had high ambitions: it was a project of a new man and a new humanity, which was to base its development on the evolution of reason, considered as a force

---

were also torn apart by the Second World War. This was a shared pool of scientific and philosophical influences, and a strong conviction of the momentous importance of information theory, statistical mechanics, cybernetics and probability, both for the sociocultural changes in the post-war world and for understanding them.

that would replace the notion of Providence, founded on religion, and perhaps even God, understood as the Supreme Reason, guiding the development of the world and giving it meaning. (2003: 48n)

According to Lem, the probabilistic revolution in science must eventually affect our understanding of humanity's place in the universe, because, seen through the prism of probability, the cosmos appears to us as a playing field. Our past is described better, more thoroughly and productively, by the theory of evolution than by creationist myths or any anthropocentric history; whereas scientists, or technicians, are better at predicting the future (or deliberately disrupting it, but that is a different story) than fortune tellers. Yet, the benefits of this shift are also anthropological. Thanks to the probabilistic cognitive framework, humankind can finally apprehend freedom without having to rely on deceptive and consolatory constructions, such as free will or Christian theodicy. In *Summa*, Lem quite bluntly condemned the finalist order of Judeo-Christian thought for its 'mechanical determinism', which leaves no room for true accidents and exceptions. Even morality in the divine order smacks of inhuman bookkeeping – all sins must be punished equally, every soul is immortal, etc. Lem preferred a probabilistic metaphysics, in which prayers are sometimes answered and one can get away with some sins. In short, it is not only knowledge that has become probabilistic: probabilistic sciences have exposed the universe as a probabilistic game. In order to orient oneself in its dizzying complexity, one must have a rational grasp of chance, abandon the deterministic and mechanistic imagination of the early Moderns, and begin to perceive the cosmos in terms of probability.

Flusser's and Lem's anthropological projects can be viewed as positive, that is, affirming the irreducible randomness of the universe and the probabilistic method of the educated guess; they are responses to a striking and liberating conclusion, formulated in 1938 by Hans Reichenbach. In his treatise on the logical foundations of science, bearing the lofty title *Experience and Prediction: An Analysis of the Foundations and the Structure of Knowledge*, Reichenbach summarised his methodical considerations about the bases and certainty of our impressions of the world with a surprisingly poetic and profound assertion:

[T]hat the relations between impressions and physical facts are probability relations, and that the certainty of the basis cannot be transferred to our knowledge of external objects. (. . .) There is no certainty at all remaining – all that we know can be maintained with probability only. There is no Archimedean point of absolute certainty left to which to attach our knowledge of the world; all we have is an elastic net of probability connections floating in open space. (192)

Reichenbach paints a picture in which the subject has no solid ground on which to tread, no absolute point of reference, but is captured in the net of uncertain relations, linking impressions and statements about the world. The message is clear but deeply ambivalent – it is not the role of a logical philosopher to attach value or existential meaning to this conclusion. Flusser and Lem both capitalise on this possibility, pursuing their respective philosophical systems, which try to familiarise this perplexing image of a groundless space of possibility.

Taming Chance with Art: An Overview

My intent in this book is to show that, even before Reichenbach's early but sophisticated reaction to the shifting status of science in the early decades of the twentieth century, avant-garde artists also responded to the groundlessness of the probabilistic world view and attempted to familiarise it. This was possible, because, as I stated above, the inevitable downfall of Newtonian science became evident with the invention of the quantum, through applying statistical methods to explain the distribution of radiation emitted by a black body. It was at this moment, at the dawn of quantum mechanics, that the Newtonian world-image was supplemented with and eventually overshadowed by the probabilistic alternative.[23] It took some time for the philosophers of science – and scientists themselves – to get their heads around the strange phenomena discovered in laboratories and the new mathematical formulae reluctantly applied to deal with them. Entire logical and metaphysical superstructures had to be carefully erected to familiarise – in sufficiently rigorous and specialised language – this unwanted intrusion of probability calculus into the very heart of science. Avant-garde artists, for their part, were quicker to react, as they were not restricted by the mandatory and time-consuming procedures of academic practice.[24] They were free to pick and choose ideas, and their means to represent them.

---

[23] In *Black-Body Theory and the Quantum Discontinuity*, Thomas Kuhn argues that Planck's innovative approach to radiation, though still grounded in classical physics, initiated a true paradigm shift by introducing the notion of the discontinuity of matter (see Kuhn 1978).

[24] Probably the first artist to notice the silent tremors emitted by the concept of 'absolute chance', causing the slow disintegration of the metaphysical edifices of Western culture, was Stephan Mallarmé, whose influential poem *A Throw of the Dice will Never Abolish Chance* finds Chance as the ultimate substitute for the compromised figures of divine infinity. It antedates Planck's physical discovery by three years. Given the enormous amount of existing literature on this topic I refrained from discussing the intricacies of Mallarmé's metaphysics and theory of poetry. For a recent interpretation see Meillassoux (2012 and 2014).

And, in contrast to the disciplined scientists and philosophers, avant-garde artists were immediately attracted to the possibility of speculation upon the existential consequences of the epistemological crisis, and eager to explore the potential for reinventing the aesthetic representations of reality it afforded.

In addition, the rebellious and experimental attitudes of the avant-garde movements made them particularly sensitive toward tropes in those branches of science which were undergoing the greatest upheavals. Revolutions in science at the turn of the century were taken at face value and eagerly welcomed as conclusive evidence that the epistemic foundations of the old culture were unfounded and detrimental to the progress of humankind; hence, new values and ideas had to be dreamt up and disseminated, among intellectual elites and the general public, as well. Their radical rejection of old conventions, which fostered static and predictable representations of reality, was tantamount to adopting an experimental attitude, consciously imitating scientific rigour and impersonal detachedness. Of course, these inspirations were sometimes adopted naïvely and superficially: in the rather liberal use of scientific and technological terms as metaphors (Max Jacob's collection of poems titled *Le laboratoire central*), in a penchant for arithmetical titles (Giacomo Balla's *Numeri innamorati*), or by inserting pseudo-mathematical formulae in poems and manifestos (Umberto Boccioni's *ABSOLUTE MOTION + RELATIVE MOTION = DYNAMISM*). However, as Renato Poggioli rightly notes, this joyful but rarely faithful embracement of modern science by the avant-garde ultimately influenced the whys and wherefores of making art: 'The laboratory and proving ground serve, in the second place (perhaps it is really the first place), an even higher aim: the technical and scientific progress of art itself' (1968: 136).

My decision to focus solely on examples from avant-garde and early neo-avant-garde art (the interwar and early post-war periods) was motivated partly by this rapprochement between art and science at the beginning of the twentieth century, resulting in an abundance of references to science in avant-garde texts and artworks, and partly by the fact the probabilistic world view, as it was concisely portrayed by Reichenbach, became fully conceivable only after 1900 and Planck's inadvertent invention. This does not mean that we cannot find other surprising and illuminating examples of the encounter between new ideas in art and probabilistic concepts in science,[25]

---

[25] One such omitted topic is the parallels between Markov's mathematical analyses of Pushkin's verse, which led to a breakthrough invention in probability calculus, and Velimir Khlebnikov's poetic and linguistic experiments. (For a historical study of Markov's work, see Link 2016.)

or that there were no attempts at 'taming chance' for creative purposes or references to probability theory and statistics in nineteenth century art.[26] Many interesting examples of the 'fabrication of accidents' are provided by Gamboni (1999), who points to isolated artistic experiments in which not only did the creative use of material randomness play a key role, but the artists themselves made tentative attempts to reflect upon their unconventional operations. In most cases, the artist's actions involved intentionally using 'accidents at work' in creative ways – taking ink blots as a source of inspiration, for example, thus ensuring that the outcome of their work was unpredictable.[27]

The most radical and unexpected instance of such an approach comes from August Strindberg, who – fifty years before Cage – experimented with a prepared instrument, a guitar he tuned at random, 'loosening the screws haphazardly', to find beautifully bizarre and unexpected chords ([1984] 1996: 104). His essay on 'the role of chance in artistic creation' is brimming with fresh and radical ideas which antedate many avant-garde practices: 'Strindberg suggested basing a new aesthetics, a "theory of automatic art," on the phenomenon of accidental images, comparing the perception of "modernist paintings," in which the viewer "assists at the birth of the picture," to the "oscillation of impressions" caused by an ambiguous natural spectacle' (Gamboni 215). However, as Gamboni cleverly points out, Strindberg and other visionaries of aesthetising randomness carried out their ideas in or referred to artistic techniques that were not their primary source of income. For example, Victor Hugo did not sign his strikingly contemporary-looking drawings with random stains of ink or coffee, probably because he feared accusations of dilettantism. Similarly, Strindberg was reluctant to randomise his dramatic output on which his livelihood depended. Chance and randomness could not penetrate the aesthetic field – even for the sole purpose of being captured, objectified and annulled – as long as art was tied to academic conventions of what forms and subjects were

---

[26] The most glaring omission in painting a historically accurate and exhaustive study of art and probability is Stéphane Mallarmé's formally groundbreaking poem *A Throw of the Dice . . .* (1897, but published posthumously according to the will of the author in 1914). As a key modernist work, it has been interpreted on numerous occasions, but more recently it has been read with a focus on its probabilistic form (Drucker 2014).

[27] It was Leonardo da Vinci who first advised painters suffering from creative block to fire their imagination by contemplating sights like 'walls spotted with stains' or 'a mixture of stones' in which a curious and frivolous eye can find unexpected silhouettes or whole scenes from which to draw inspiration (da Vinci 2008: 173).

appropriate (nineteenth-century artistic professionalism ruled out a reliance on chance).[28]

Probability and statistics appear in nineteenth-century visual arts less directly as well, as an afterthought of the discretisation of the image by painters like Georges Seurat, Paul Signac or Camille Pissarro. Their pointillist technique of building an image from tiny dots of pure pigments instead of blending them on a palette, to trick the eye into seeing colours, tints and shades, was based on scientific studies of colour perception. Many scholars have noticed that this technique anticipated modern imaging technologies, like colour printers and digital screens, which also rely on dividing the display surface into tiny, discrete units (Damisch 2007). Conversely, it has also been argued that pointillist painters were responding to the existing photo-mechanical printing technologies, which also operated by breaking up the image into thousands of smaller units (Broude 1974). Whichever viewpoint we take, the fact is that pointillism introduced the principle of physical discontinuity into art history, as pointillist painters populated canvases with (almost) standardised tiny elements – large assemblies of pictorial atoms. Thus, pointillist technique lends itself to the possibility of being read as a kind of statistical sensibility: seeing the macroscopic world as an emergent product of interactions between microscopic entities too tiny to be seen by the naked eye. There is, however, a crucial difference between the statistical sensibility of artists like Seurat or Signac and those discussed in my book. To pin this down, and simultaneously disclose a crucial characteristic of all the examples collected in this book, I would like to juxtapose two pictures, both rendered in similar form and both intended as a social commentary: Seurat's *A Sunday Afternoon on the Island of La Grande Jatte* (1886) and Boccioni's *The City Rises* (1910).

In Seurat's *La Grand Jatte*, a statistical sensibility shines through, first and foremost, its pointillist form: the image is composed of tiny dots and brushstrokes which remain visually distinct despite being united by the human eye to create tones, shades and shapes. From these tiny particles of paint, Seurat assembles a satirical image of

---

[28] The first art school to systematically embrace chance operations was Black Mountain Collage (opened in 1933 and closed in 1957). Founded on the multidisciplinary tradition of Bauhaus and directed in the spirit of John Dewey's philosophy of experimentalism, it hosted courses by Buckminster Fuller and John Cage who were open to open-minded collaboration which became more important than simply imparting knowledge and skill to students (for a comprehensive overview of their different methods see Díaz 2015: 53–148).

French society: a gallery of distinct types, lacking neither individuality nor substance, coexisting in motionless apathy in a place of leisure on a riverbank on Sunday. Everyone seems to have escaped there, to this peaceful place on the outskirts of Paris, and seem equally dazed and bored by their tiresome existence in a modern metropolis. Their coexistence, however, is accidental or even outright improbable, as T. J. Clark has noted, as it seems unlikely that workers would intermingle so closely with the bourgeoisie (1984: 262n). Jonathan Crary rightly asks if this coexistence of different classes should imply that the image depicts a state of social harmony, or rather, 'is it a statistical distribution of isolated and categorized units, the result of a merely additive principle of formal adjacency, in which depleted, anomic relations predominate beneath the spurious appearance of social concord?' (2006: 179). It is difficult to ascertain if Seurat remains affirmative or critical of his own point of view – if *La Grand Jatte* is to be read as an affirmation of a mass capitalist, but increasingly democratic, society which appears to erase rigid social distinctions, or whether it critiques the statistical world view in which reality appears discontinuous, uncertain, and threatens individuality. For Clark, however, neither of these answers is correct. Rather, it is a failed attempt at a class critique; an example of a bourgeois, ideological view of a society supposedly without class distinctions, where only superficial differences in appearances remain (266n). Whichever version is true, one thing remains certain: the society is depicted in utter stasis, as there is no movement. It is also statistical, as not only are the figures randomly distributed in space, they are also assembled from tiny atoms of paint.

The latter can also be said of Boccioni's first Futurist painting, *The City Rises*, exhibited over two decades later, although the particles do not hang uniform and motionless in space, they are smeared irregularly on the surface. Furthermore, here we see a different social situation: a crowd of workers and horses moving feverishly about a construction site. In contrast to the dreamy atmosphere of a riverbank in Seurat's picture, Boccioni depicts the city frantically and enthusiastically expanding, devouring peaceful places of respite. Seurat's ambiguity and irony are completely absent here. There are no individuals, or even classes, only a messy assemblage of tools and bodies, both human and animal. Distinct shapes dissolve and liquefy to form a chaotic, indeterminate vortex of boiling matter, transforming figures into intense quasi-abstractions. As the city rises, it also descends into ecstatic chaos and spirals out of control. The visual effect of the dissipating bodies derives from the divisionist technique of using only tiny strokes of paint, making the transitions between bodies and their surroundings hazy and uncertain. People and animals seem to be extended in space, though not in any precise, determinate direction, as in Eadweard Muybridge's iconic photographs of dissected motion. Here, figures expand and merge with one another, as this intricate and chaotic physical system of workers, tools and animals evolves into an uncertain future(s). If in Seurat's picture human figures resemble a 'statistical distribution of units', Boccioni's figures quite literally dissolve into waves (of probability) smeared onto the canvas. If the former depicts objects frozen in space, the latter tries to capture a dynamic and non-linear system.

This technique eventually became a hallmark of Boccioni's style, representing potential movement by superimposing different pathways (or futures) of a figure in motion. An even better and more vivid example of this technique is *Dynamism of a Football Player* (1913), a study of a single figure which borders on complete abstraction, as only the title and a single calf sticking out of a whirl of colour reveal the painting's subject. The footballer faces the viewer, who is put in the position of a bewildered defender trying to assess and anticipate his opponent's possible moves, suspended in the state of possibility. Boccioni tries to capture these possible futures – various paths of movement – and represent them simultaneously. What we see is not a player in motion, but the perception of his movement, simultaneously seen and inferred. The painting documents the act of looking as a probabilistic operation of a mind struggling to guess what could happen next. Although the uncertainty of perceived motion is captured in a static form, and thus tamed, it is never dispelled. On the contrary, it occupies the centre stage, as it is investigated, celebrated and aestheticised. In a sense – and this remark pertains to all the examples gathered in this book – uncertainty and randomness have been tamed, aesthetically and not mathematically, of course, but at the same time they have finally become expressible.

The history of probability theory shows that mathematical mastery of chance eventually led to acknowledging uncertainty and chance as inescapable elements of thought and of reality itself. Only after the evolution of statistical mechanics and the emergence of quantum physics did science develop probabilistic tools not only to tame chance, but also to objectify its indeterminate nature. Similarly, many avant-garde artists, in trying to tame the untameable, devised innovative, experimental ways of representing reality and making art: various configurations of probabilistic aesthetics intended to sensitise the modern human to the randomness at the heart of matter and the inevitable uncertainty of our perceptions.

In Chapter 2, I look at the Futurist sensibility and investigate its indebtedness to the emerging discourses of post-Newtonian science. In a late Futurist manifesto, 'Qualitative Imaginative Futurist Mathematics' (1941), Filippo Tommaso Marinetti, Pino Masnata and Marcello Puma proclaim: 'Let each person make his own subjective calculus of probabilities.' In this fascinating piece, affirming 'the divine essence of CHANCE and RANDOMNESS', the Futurists explicitly articulate their undying interest in the probabilistic concepts of natural phenomena. I maintain that a conviction of the inherently probabilistic nature of perception and cognition, a claim incompatible with the rigid machinism usually associated with Futurist aesthetics, was at the heart of many of their earlier artistic explorations. To back this claim, I focus on three distinct cases: Rosa Rosà's *A Woman with Three Souls*, read as a manifesto of the science of the future; Umberto Boccioni's theory and paintings of 'states of mind', indirectly informed by Hermann von Helmholtz's concept of unconscious inferences; and Bruno Corra and Emilio Settimelli's manifesto about the prices and values of artistic genius, inspired by associationist theories of the mind and Georg Simmel's treatise on the cultural and mental effects of capitalism. In investigating the concept of the Futurist sensibility, I aim to show how this Italian movement forged revolutionary forms of existence, an uncannily visionary understanding of matter, and new ways of perceiving the world, revolving around the very act of prediction. To conclude, I discuss the Futurists' ideas in relation to Bruno de Finetti's philosophy of probabilism, developed in 1931 and directly inspired by avant-garde art and revolutionary political movements in Italy. Assuming that probabilism, alongside 'activism, relativism, fascism, futurism, bolshevism', are 'different aspects of a single reality, of which we are all children', de Finetti considered all knowledge produced in the twentieth century to be ultimately subjective, conventional, operational and uncertain.

The third chapter opens with a sudden change of setting, but returns to familiar territories of the avant-garde in the latter part. I begin by presenting three theories of the mind that share some key probabilistic assumptions and were conceived around the same time, in the late 1940s and early 1950s: Warren McCulloch and Walter Pitts's model of neural networks, Friedrich Hayek's book on the neural organisation of the human sensory domain, and Jacques Lacan's incorporation of the probabilistic epistemology of cybernetics into psychoanalysis. I argue that these three approaches supplement each other in crucial ways, presenting the idea of probabilistic thinking from different points of view: mathematical (McCulloch and Pitts), epistemological/economic (Hayek) and aesthetic/existential (Lacan). Discussing Lacan's case, I show his direct indebtedness to surrealism and go on to discuss the probabilistic ideas in the theoretical writings of Salvador Dalí and André Breton on the unconscious. The former's zany theories of the human mind display the fascinating self-awareness of an artist who defined success in terms of mastering improbable associations, remote from common sense, consciously drawn from the vast and random fields of the unconscious. On the other hand, Breton takes a different position, as he wants to observe, study and aestheticise the random expressions of the unconscious. I compare his methods with the early practices of randomisation in laboratories and show how this approach was informed by early uses of probability calculus in psychology.

In Chapter 4, I investigate the role of probabilistic intuition in the highly abstract approach to art practised by Duchamp, who, despite being loosely associated with surrealism (and greatly admired by Breton), ventured on a unique and solitary intellectual path. There is no doubt that Duchamp was keenly interested in the notion of probability: he left perplexing comments in his notebook to *The Large Glass*, which include many reflections on chance and possibility; he printed bonds for his system of playing roulette; and he even described one of his works, *Three Standard Stoppages*, as 'canned chance'. After analysing these examples, I turn to Duchamp's ideas on art, in particular, the concept of the 'figuration of a possible' and his view of art as an intellectual game with the viewer's expectations. Finally, I discuss how these ideas were connected to Duchamp's impersonal theory of creativity and his unconditional appreciation of freedom from the boundaries of a fixed identity. To elaborate on this unlikely point of view, I demonstrate how Duchamp's radical stance is embodied by Ulrich, the main character in Robert Musil's *Man without Qualities*. Both characters blur the distinction between truth and fiction and condone a kind of non-subjectivity, in which the notion of possibility becomes crucial. I take a closer look at their existential experiments, fuelled by randomness and designed to reach the farthest boundaries of improbability.

Chapter 5 is devoted to the art and philosophy of Duchamp's intellectual heir, John Cage. In one of his most influential lectures on indeterminacy, *Composition as Process*, he compared a musical composition to a camera and implied that performance could be conceived as the act of taking photos. The idea of composition-as-camera suggests a certain understanding of the materiality of sound. Cage's theory of composition and indeterminate performance responded to the emergence of a new epistemology of information and conveyed a probabilistic agenda: a way of apprehending reality that only became possible after the first computers were up and running. Cage's novel approach to performance is not only an important example of an intellectual affinity between neo-avant-garde artists and cyberneticians; it also proves that, long before these groups started collaborating in New York, they shared an anti-representational stance. Moreover, we might argue it was the probabilistic epistemology of cybernetics that decisively informed and shaped the performative turn in the United States. As I argue, Cage was well aware that the relation between indeterminate performance and technological reproduction was not conflictual – on the contrary, it was unavoidable. Indeed, it was the possibility to capture vibrations of matter and translate them into machinic codes that allowed him to appreciate noise as a resource in art. Cage's path to 'nature in her manner of operation' inevitably led him through alienation and into technological and probabilistic abstraction. Impersonal and faceless technology restricts and confines the egotistical drives of the artist, creating a safe space, free from individual tastes, social bias, cultural backgrounds, etc. The aesthetic pleasure Cage projected hinges on the ability to enjoy unpredictable events that defy expectations, which is only possible in a secure and non-hierarchical environment. His aesthetic (and political) heterotopia thus had an important probabilistic dimension: to be made possible and avoid cultural conflicts, indeterminate performance implies that everyone perceives the world in terms of shifting probabilities; despite their differences, everyone absorbs incoming data to make predictions (in contrast, not everyone thinks logically, speaks Spanish or finds pleasure in melody). The idea of indeterminate performance, in which surprise is celebrated in and of itself, is contingent on this (non-)ideological assumption.

In the sixth chapter, I discuss the compositions and philosophical writings of Iannis Xenakis, who began developing his own method of 'taming chance' around the same time as Cage, but with different results. Of all art projects analysed in this book, the music composed by Xenakis in the late 1950s and early 1960s was most closely linked to the mathematical theory of probability. After graduating in engineering and spending years as an assistant in Le Corbusier's studio, he developed his signature sound by applying mathematical tools – such as statistical mechanics, Markov chains and normal distribution – in the creative process. Olivier Messiaen, Xenakis's mentor, considered his pupil's music to be radically different from anything that could be heard at concert halls at the time. It did not belong to any trend in classical music, but what many critics found the most striking was its ecstatic and yet inhuman quality. In my reading of Xenakis's work, I focus on the role and significance of probability theory, though in strictly philosophical terms. I argue that, not unlike the Futurists, Xenakis wanted to bring about a new kind of 'probabilistic sensibility' that derived (nihilistic) pleasure in finding a structure in a sonic mass.

Moreover, drawing inspiration from Presocratic philosophy and historical analyses of ancient music and rituals in Greece, he attempted to give a different meaning to music, which I characterise as a sonic representation of dzoē, a form of undead existence, beyond the distinction between nature and artifice. To demonstrate the particularity of Xenakis's aesthetics and philosophy, I will compare his idea with more technically inclined theories of computer art from around the same time, whose proponents made use of probability theory without assigning it such existential significance.

Art and Science: Troubled Relationships

To sum up, all the chapters collected in this book provide a non-linear history of what I am calling probabilistic sensibilities: novel modes of sensing and making sense of reality, developed throughout the twentieth century by various avant-garde experimenters. In every case, I demonstrate the connections between the artworks, the manifestos accompanying them and probabilistic notions from scientific discourse. These investigations into epistemology and aesthetics are based on the fairly uncontroversial assumption that scientific knowledge and experimental art differ in principle but nonetheless share some important features. In the words of Jacques Ellul:

> Heuristic strategies, the direct examination of experience, the positing of more-or-less-fruitful hypotheses – art and science advance side by side, in defiance of dogma and in quest of the experience that can legitimate them. Science enables us to restructure the external world, art to interpret our experience of that world. (2001: 26)

Western artists, especially since the shift toward experimentation at the turn of the centuries, have tried to provide the public with structures through which they can experience reality in a novel way – training their eyes, ears and noses to take in stimuli and organise them differently. If by knowing we simply grasp a form of structured awareness which allows us to make sense of our environment and predict its changes, then science and art both offer such operational structures.

Besides straightforward cases that thematise the probabilistic aspects of cognition and perception in artworks and essays, I also call attention to strategies of experimenting with chance: more or less rigorous procedures of controlling randomness – 'chance operations'. The term, borrowed from Duchamp's notes, implies a procedural attitude toward making art: operating on chance involves defining a set of rules, or 'a field of preestablished determinations', mimicking, to some extent, laboratory practices (see Kepes 1969: 194–6). In this sense, certain musical compositions by Cage can be compared to attempts to 'tame chance' with probability calculus, because the rigid frame constructed by the composer determines behaviour and creates an environment where chance events take centre stage and the inferential nature of perception is pronounced. Sometimes these procedures were executed laboriously, sometimes gaps and distortions were introduced to intentionally allow uncertainty and randomness to leave its mark on the creative process, disassociating the artist from their creation. Either way, for all the artists mentioned in this book, these sets of rules were important enough to be included in long manuals, in the form of manifestos explaining how their art came to be or how it should be handled. I pay so much attention to these textbooks as I am seeking direct or indirect links between the artist's self-reflexive remarks and probabilistic concepts in various fields of science.

To shed some light on the probabilistic logic behind these artistic and theoretical projects, I refer to concepts that both pre- and antedate them. For example, writing on Cage's art, I invoke concepts developed by cyberneticians in the 1940s, which were undoubtedly known to the composer, and more recent theories of cognition. In doing so, it is not my intention to present avant-garde artists as prophets who dared to state hypotheses on the nature of perception that would not be developed into widely accepted scientific theories until decades later. Rather, following Flusser and Lem, who argued that the notion of probability was crucial to understanding the logic of a global, digitally connected world, I want to call attention to the aesthetic (senses-related) and experiential dimensions of the probabilistic world-image. Besides the historical and theoretical implications of unearthing the avant-garde's indebtedness to probabilistic ideas in science, there are also more pressing and practical reasons for my undertaking. Presenting probability in the context of art and aesthetics, remote from the confines of mathematical symbolisms, seems to me particularly important at this moment, as we are increasingly surrounded by probabilistic bots, and the very nature of cognition is being reconceived as inherently probabilistic. These elusive AIs – bots designed to predict what we want to write next (digital writing assistants) or what we would like to buy (algorithms on shopping sites) – have already penetrated our mental environments enough to radically change our practices in using digital platforms. Their seamless integration into everyday activities – communicating, searching for information, shopping, listening to music, and so on – hinges precisely on their probabilistic logic, which allows them to function among human beings in conditions of complexity and unavoidable uncertainty. Indirectly, their success story as a companion species tells us something about ourselves – we speak a common language of prediction.[29] The notion of probability is thus becoming increasingly useful in understanding the logic of the globalised world as it is transformed by information technologies which are permeating almost every aspect of our existence and replacing old cultural systems of communication run on out-of-date software (on truth, meaning and – horror of horrors – certainties). To make more sense of this new situation it might be useful to grasp how the probabilistic world-image has manifested itself in experimental art and has become assimilated by artists. In other words, exploring the probabilistic 'operations' of avant-garde art may help to understand how information technologies challenge and shift collective patterns of perception and cognition.

All analysed cases – except the grand audio-visual projects by Cage and Xenakis carried out in the 1960s – belong to the period when artists and scientists often exchanged and responded to each other's ideas, yet indirectly: through publicly exhibited artworks or essays published in publicly available sources. Before the Second World War they rarely, if ever, undertook any joint projects. For the members of most avant-garde movements in Europe at the time, especially the Italian Futurists, the German Dadaists and the French surrealists, science was attractive but only as a discourse to be observed from a distance and exploited in search of novel concepts, or work methods which could have been mimicked in their practice, often in a parodic or ironic tone. The zany and erratic cultures of the avant-garde and the serious scientific societies could not get along easily, not only because of the insurmountable differences in character but also due to the lack of common practical goals, or even shared attitudes toward social changes brought about by scientific and technological progress.

However, this condition of genuine separation between art and science changed soon after the Second World War and three factors played a crucial role in the process. Firstly, on the level of material conditions, the beginning of the Cold War marks a period in which Western universities, especially in the United States, received huge injections of funds, both from public and private sources, which in the process opened some possibilities for launching experimental collaborative projects, fostered by institutions like the Center for Advanced Visual Studies at MIT or Experiments in Art and Technology (E.A.T.) formed by engineers associated with Bell Labs

---

[29] This is best evidenced by the extent to which, in recent years, 'AI and neuroscience drive each other forwards' (Savage 2019).

(both established in 1967). Secondly, the pressure of the Cold War and the rapid expansion of the US economy created a growing need to innovate – deliver new ideas and solutions to social issues – bringing artists, scientists and engineers closer together in their pursuit of the unforeseen.[30] Thirdly, for some visionaries, like Buckminster Fuller and Billy Klüver, the founder of E.A.T., a collaboration between all three professions promised a way out of alienation, instigated by professional specialisation and compartmentalisation of society into prefabricated and rigid roles in the workplace, and offered a glimpse into the possible future of democratic society, transformed by knowledge, technology, collaborative work and creative freedom associated with the figure of the modern artist. Close encounters between artists and scientists, taking place since the 1950s in universities and corporate spaces, furnished with cutting-edge equipment (from the first computers to lasers), came at a price, though. These spaces were funded primarily by institutions belonging to the so-called 'military-industrial complex', or private corporations, which allocated their resources with specific intentions in mind, and which were often at odds, to say the least, with the political and ideological inclinations of the avant-garde and neo-avant-garde artists. It is thus true, as John Beck and Ryan Bishop remark, that through such epistemic connectors, like the notion of probability, art does not connect exclusively with abstract scientific knowledge, but also the uncomfortable baggage of material, historical practices and events made possible thanks to those seemingly abstract concepts:

> The work on probability that made possible the innovations of the Manhattan Project and migrated, postwar, to think tanks like the RAND Corporation, where game theory underpinned nuclear strategy before finding widespread applications in economics and the social sciences, is a narrative that runs alongside, and periodically intersects with, the history of the artistic avant-garde. This dual track, crisscrossing series of knots and loops binds the history of art and the history of science and technology in the US during the twentieth century. (Beck and Bishop 2020: 15)

Probably, the wide-spread popularisation of probabilistic tools for automated creativity opens a fresh chapter in the shared history of mathematical probability and (avant-garde once again?) art. But this is a topic for another book.

---

[30] On relations between aesthetics and post-war American technocracy see Lee 2020.

Chapter 2

The Futurists: Hooked on Uncertainty

In 1931 Burrhus Frederic Skinner, a founding father of behaviourist psychology, began his research into different forms of a simple reflex – he spent over fifteen years in a lab with rats, pulling levers, delving into a seemingly simple question: Can cognition be explained in determinist terms as causal chains of action and reaction? His initial intention was to map 'relationships within a regressive series of events extending from the behavior itself to those energy changes at the periphery which we designate as stimuli' ([1931] 1999: 446). Eventually, among all these simple 'regressive' processes that proceeded from physical events to neural excitement – from levers being pulled to dopamine being released – one particular class of behaviour captivated his attention the most: the initial pull. What remained so puzzling in this simple activity was the fact that he could come up with no causal explanation for it. When the animal establishes a link between an activity and its effect leading to neurochemical rewards, the causal nature of this process is usually self-evident. Yet this seemed like a rat's 'leap of faith'. The causal interpretation cannot apply to a pull that has never rewarded (or suggested by rat elders). A rat which has never pulled a lever in its lifetime cannot infer that 'levers + push = food'. It took Skinner over twenty years to admit that some behaviours cannot be causally explained. After two decades of watching generation after generation of rodents gratuitously pulling levers he accepted that some behaviours – primarily those related to the search for food and exploration – can be 'randomly emitted'.

In an article titled 'Probabilistic Thinking and the Fight against Subjectivity', Gerd Gigerenzer recalls this story to show how painstakingly difficult it was to convince academic psychologists of the probabilistic nature of cognition. And even if they did make use of probabilistic models, it was only on the methodological level of 'measurement and data description'. Gigerenzer finds it baffling that

> [b]ound to determinism and objectivity as their ideals of science, experimental psychologists dealt with probabilistic thinking in quite a different way than their physicist contemporaries [. . .]. The irony is that the harder sciences, like physics and biology, introduce uncertainty via probabilistic models into their subject matter, whereas the so-called soft ones, like psychology, were loath to tolerate such uncertainty, and long clung to the ideals of classical physics. (1987: 29)

He concludes that ever since the Enlightenment, determinist beliefs had been considered necessary for scientists to be acknowledged as true scholars. Psychologists' efforts to find deterministic laws were motivated by a powerful need to be taken seriously by their scientific peers.

Having no reputation to lose, it was the playful and buffoonish avant-gardists who first developed the idea that cognition might be a probabilistic activity, and even went on to speculate on the epistemological ramifications of this strange contention. After rejecting the all-too-serious culture of the bourgeoisie, which left them no room for free exploration and experimentation within its professionalised ranks, they began to play rebelliously with whatever lever they liked in – or against all – hope of unexpected rewards falling from above. Paradoxically, among all avant-garde movements at the beginning of the twentieth century, it was the Futurists who became fully aware of their role in the evolution of society: as explorers trying to inhabit new territories, where the notions and values of the old culture no longer applied. Looking at their paintings, numerous manifestos and literary works – primarily those by Filippo Tommaso Marinetti, Umberto Boccioni, Bruno Corra, Emilio Sentimelli and Rosa Rosà – one glimpses numerous sparks of a probabilistic self-awareness. All of them – though each in a slightly different way – attempted to understand and lay the foundations of their newly invented identities, which resolved around the essentially probabilistic act of predicting the future.

By virtue of their auto-reflexive and theoretical preoccupations, the Futurists were naturally inclined to experiment with probabilistic motifs, such as chance, randomness, noise, improbability or prediction. And although they shared a critical distrust toward the science practised by the academic community and often wandered into obscure intellectual territories of the occult, so popular at the turn of the century, they were also very keen to build upon scientific concepts they dug up in newspapers, magazines and popular science books. Among many others, two sources in particular resonated with the Futurists, providing the movement with touchstones for their provocative intellectual speculations. The first was Gustave Le Bon's *The Evolution of Matter*, a handy collection of appealing, ready-to-use concepts revolving around the dissociation of matter, which was for a long time guarded by the dogma of indestructibility. Invoking discoveries and inventions such as X-rays, cathode rays and radioactivity, Le Bon argues that modern physicists challenge the notion of matter as essentially indivisible, fixed and stable. This concept was of prime importance to Futurist painters in search of a visual language to depict not only the dizzying speed of modern vehicles, but also the microscopic movements of matter hidden from the naked eye (Le Bon 1909). The other book on the Futurist shortlist of must-read popular science works, *Science and Hypothesis* by Henri Poincaré, was one that also had a major effect on Le Bon himself, who misappropriated and exaggerated some of Poincaré's points. Poincaré, a renowned mathematician, physicist and engineer, unlike Le Bon, had published this collection of essays on a seemingly obscure topic in the philosophy of science – on the special status of certain physical theories. However, his radical and seditious treatment of the subject caught the interest of both the scientific community and public opinion. Poincaré's central claim that caught Futurists' attention and served as a vantage point for further exaggerations and speculations was that many physical theories are founded on conventions that are useful but not necessarily true. A great many scientific theories, he argued, do not meet the criteria of universal truths and should be considered probabilistic estimates. Poincaré reasoned that scientific problems are not that different from everyday efforts to find sufficient criteria for action in a state of uncertainty. In Poincaré's work, Futurists found a voice of authority who questioned the dogmas of positivism and Newtonian science and incidentally became exposed to the idea that linked serious, epistemological considerations about the probabilistic nature of scientific reasoning with more general, almost existential problematics (Poincaré 2018: 129). We will engage in depth with Poincaré's thought in the next chapter, partly because its protagonist, Marcel Duchamp, was engaged in a more nuanced and direct debate with the French polymath, and, secondly, the Futurists in general tried to avoid referring directly to figures of established authority.

The Futurists' rejection of hard science originated from a strong feeling of disdain for the academy, which they perceived as an ultimately conservative and bourgeois institution that only held back the evolution of societies instead of actively encouraging their trans-formation.[1] For instance, the first on the list of old institutions on the brink of revolutionary transformation listed in Corra and Setti-melli's manifesto 'Weights, Measures, and Prices of Artistic Genius' (1914), are 'the laboratories and schools of passéist science', not the galleries or museums (2009: 186). The artists were marching one step behind the real avant-garde of scientists, inventors and industrialists. Even though the Futurists were never keen on under-estimating their own importance, they also universally agreed that Futurism was made possible in the first place by – as they state in 'Manifesto of the Futurist Painters' (1910) – the 'triumphant prog-ress of science' which brought about 'changes in humanity so pro-found as to dig an abyss between the docile slaves of the past and us who are free' (Boccioni et al. 2009: 62). Furthermore, the Futurists rejected the Romantic ideal of an artist as an outlaw, only creating art 'for art's sake' (a notion still very much alive among other avant-gardists), replacing it with a new model: a socially useful worker whose products, as Corra and Settimelli bizarrely insisted, can be pre-cisely measured and properly remunerated. Unlike industrial work-ers, his performance would be evaluated not in terms of quantity, but relative unpredictability. The Futurist artist was thus supposed to produce the exceptional – the unheard, unseen, unfelt – instead of the merely useful. Yet this still meant that pulling levers randomly in the laboratory of 'intoxicating modernity of contemporary life', as Boccioni defined his 'field of research' in 'The Plastic Foundations of Futurist Sculpture and Painting' (2009: 142), was considered a practical occupation deserving of social approval and legiti-mate reward. For this to occur, however, the artist had to team up with mathematicians and engineers to create the infrastructure for measuring genius. This idea predates the arrival of the so-called 'creative class' in its contemporary understanding: a group of people driven by individual profit and rare achievements, who nonethe-less want to inhabit the mainstream of social and economic life (see Florida 2002).

This chapter is devoted to the work of a relatively small group of artists and philosophical dilettantes in Italy who, on many lev-els, became aware of the numerical aspect of human existence and the alienation afforded by drastic changes in living conditions in the modern world. Instead of revolting against this new situation in the name of old values, these remarkable eulogists of change tried to

---

[1] Many claim otherwise – that Futurism was Bergsonian and non-positivist.

familiarise this growing sense of alienation by embracing the feeling of uncertainty and consciously aiming for improbable ideas and biographies. However, in their pursuits, they still strove for the wider social acceptance and understanding of the social elites. For that reason, they were bound to explain themselves as clearly as possible, and theorise about their (im)probabilistic inclinations. As Jeffrey T. Schnapp has noted:

> Global Futurism (. . .) sing[s] the statistical battlefields of modern war, the city's number-studded landscapes, and a world of human multitudes navigating a sea of that newest product of contemporary social sciences: statistics regarding demographics, production, consumption, and mobility. And these worlds of number are closely allied with and indeed animated by, on the one side, mathematics as associated with contemporary technics and, on the other, statistical hype of the sort extending from commercial advertising to political propaganda. (2012: 109)

Schnapp describes the aesthetic ambition of Futurism as the 'statistical sublime' – a fascination with mundane data, nonsensical equations, never-ending inventories of objects, and random numbers. I want to supplement his theory, focused solely on the poetry, with an analysis of the Futurists' philosophical positions. In my attempt at an overall mapping of their probabilistic inclinations, I will focus on three topics: (1) their embrace of post-Newtonian science; (2) the concept of Futurist sensibility and Boccioni's theory of 'states of mind', indirectly informed by Hermann von Helmholtz's concept of unconscious inferences; and (3) Corra and Settimelli's manifesto on the prices and value of artistic genius, inspired by associationist theories of the mind and Georg Simmel's treatise on the cultural and psychological effects of capitalism.

It is no coincidence that, in the eyes of Bruno de Finetti, one of the last century's foremost philosophers of probability, Futurism was one of those intellectual currents that signalled fundamental socio-epistemic shifts: from an (objectivist and universalist) fixation on truth and firm metaphysical foundations toward a (relativist) acceptance of probability as the new ultimate standard for knowledge. In 'Probabilism', first published in 1931, a lengthy essay / manifesto on the epistemological significance of probability in science and beyond, de Finetti concludes that his philosophy of probability belongs to the same group as '*activism, relativism, fascism, futurism, bolshevism*', which are merely different aspects of the same civilisational change (218). Although this grouping of intellectual and political movements initially seems fundamentally incoherent, de Finetti believed, like the

Futurists, that the fascist spirit of revolutionary action was fuelled by contempt for objective truths and fixed values (see Galovotti 2008). And again like the Futurists, he combined praise for Mussolini's collectivist policies (instead of boundless faith in the social equilibrium engendered by the free market and democracy) with anarchism in such matters as morality. These radical political views were consistent with his subjectivist interpretation of probability as a level of belief which only need obey the operational and pragmatic requirements of formal coherence. De Finetti rejected objective reality as an unnecessary metaphysical supposition of little operational and pragmatic value:

> So no science will permit us say: this fact will come about, it will be thus and so because it follows from a certain law, and that law is an absolute truth. (. . .) What we can say is this: I foresee that such a fact will come about, and that it will happen in such a way, because past experience and its scientific elaboration by human thought make this forecast seem reasonable to me. (170)

In the twentieth century, he argued, knowledge should be reframed as a concrete product of human activity to find one's bearings in one's environment by making predictions. His logico-mathematical theory of predictive inference followed this foundational statement and, as he was unafraid to admit, was not the product of a lone scientist, but expressed the sociopolitical climate in Europe after the First World War. Such sentiments and convictions like the rejection of truth, radical individualism and a simultaneous interest in the materiality and inferential nature of perception found their place in the seemingly remote discourses of science and avant-garde art.

The Uncanny Physics of Futurism

There are multiple reasons why Rosa Rosà's *A Woman with Three Souls* (1918) is a prime example in beginning our investigation of the complexity of the Futurist attitude toward science: it is primarily concerned with profound social impact of scientific innovation; it criticises the mediocrity of academia and pays tribute to the unknowable future of science; and finally, in its rejection of philosophical determinism, it announces the probabilistic revolution. Rosà, whose real name was Edyth von Haynau, was born in Vienna in 1884, and became involved with the Futurist movement during the First World War. Her short science-fiction story, written in 1918, is at face value a feminist parable and 'a manifesto about the impending transformation of women's lives, personalities, and gender roles in the twentieth century and beyond' (Re 2011: 2). Yet if it is read as a treatise on the social and psychological effects of science in the future, it reveals a complexity far greater than its feminist message. Rosà tells the story of the miraculous transformations of Giorgina Rossi, a character completely devoid of any characteristics or exceptional traits – an illustration of a statistical *femme moyen*. Only as a result of an accident in a laboratory next to her apartment does she have the opportunity to experience brief moments of superhuman exceptionality. In each of these moments she is transformed into a different (super)woman from the future: an erotic *flâneuse* who transgresses the moral codes of the patriarchy, a public intellectual who, like an automaton, rattles off one scientific theory after another to a bewildered crowd, and finally, a spiritual poet who dreams of leaving her body and drifting away into the cosmos. Of course, all three metamorphoses carry symbolic undertones. On the surface, *A Woman with Three Souls* offers a glimpse into the future of the feminist movement. The story can be read as a prediction of three distinct stages of a feminist and, to a lesser extent, transhuman social liberation – moral, intellectual and creative – as the second and third transformations involve both empowerment and dehumanisation. For instance, as Giorgina becomes possessed by scientific discourse, she is reduced to a medium through which the abstract and mathematical is embodied and expressed. The same drive toward abstraction is the theme of the last metamorphosis, which suddenly occurs when Giorgina is in the middle of writing an ordinary letter to her husband, but is distracted by a poetic revelation. Everyday matters give way to part-mystical, part-scientific fantasy:

> I travel through spaces filled with billions of astral storms, incredibly violent electric charges. [. . .] In my meteoric trajectory, I feverishly pierce through space, travel through all its dimensions, creating new forms, liquid combinations, abstract shapes, blinding hot energy. In my flight I trace astral curves that appear with a tawny hiss against the backdrop of the Infinite. (2011: 34)

Not coincidentally, Giorgina's revelation leads her to fantasise about losing her identity in cosmic swirls of matter and energy. Rosà, who in her lifetime oddly identified both with the feminists and the often misogynistic Futurists, creates a sociopolitical parable which transcends the ideological limitations of both movements. Her vision of the feminist future is less about political powers, or social status, and more about the effects of science and technology exerted on the women of the future.

Rosà makes it explicit that Giorgina does not owe her enlighten-ment to herself – or any other woman, for that matter – but to a scientific accident which happened completely independently of her volition: to 'splinters of time' which 'materialized in our era' after an electrical storm interfered with the scientific equipment in a labora-tory. This suggests that Rosà saw cultural and political transforma-tions as consequential to essentially contingent events, taking place beyond the scope of human control, provoked and described by sci-entists, domesticated by artists, and translated into political move-ments by social activists (but always *post factum*). On top of that, her depiction of science, particularly of its partly accidental progress, displays a surprisingly Kuhnian understanding of scientific practice. In his seminal *The Structure of Scientific Revolutions* (1962), Thomas Kuhn argued that, every once in while, a sudden event – an inven-tion or a discovery that alters the very terms and conditions of what is considered knowledge – leads to a 'paradigm shift'. The nature of the research prompting such sudden revolutions must be exploratory and deal with anomalies that cannot be explained in terms of the status quo: contrary to present-day wishful thinking, true innovation cannot be programmed. The electrical discharge that interacts with lab equipment to trigger the fragmentation of time is an allegory for anomalies that push science forward.

In Rosà's novel, the scientist, but also the artist herself, profit from such disturbances. As an author, Rosà comes up with unusual events in the complex fabric of 'contemporary life' and creates models of existence and sensibilities for others to inhabit. An anomaly disrupt-ing the fabric of contemporary life allows her to anchor her imagi-nation and freely speculate about possible futures. The character of Giorgina can be interpreted non-symbolically as a dummy, devoid of free will or agency, and used instrumentally by Rosà to present various 'techno-feminist' futures. In turn, the anomalous 'electri-cal charge' would be a singular event that opens a window to the possible future(s). Beyond any doubt, Rosà, alongside many other Futurists, had a knack for spotting and capturing the distant patterns through that window.

The most mind-boggling example of Rosà's fortune-telling prowess concerns the various effects of the techno-scientific anomaly on men and women. The only scientist affected by the event, like Georgina, turns into a vessel of time, though in his case the transformation takes on a different form: he begins to age rapidly, as his hair and nails grow to outrageous lengths. Rosà implies that if women have a bright future ahead of them – as femme fatales, brilliant scholars and inspired art-ists – then men are doomed to devolve into primitive and grotesque cavemen. This could be read as a bold prediction of the masculin-ity crisis (Horrocks 1994). Another glimpse into what may lie ahead for modern civilisation concerns fluid (interchangeable) identity. Topics such as identity theft, somnambulism and possession were often addressed by the artists in the first half of the century, for example, by German Expressionist filmmakers or French surrealists. However, they were rarely depicted as positively as in Rosà's novel. Even the scientists, who initially treat poor Giorgina with electro-shocks to prevent future attacks of 'hysteria', accept the fact that

> [t]he coexistence of multiple personalities within the same organism will bring an end to the continuity of consciousness and, consequently, to all ethical and legal responsibilities. Since this future is probably closer than we think, it will be necessary to prepare for a complete change in all the moral and legal codes that have regulated our Society thus far. (38)

Speaking through the mouths of the old scientists, Rosà argues that personality need not be considered an essential core of human psyche but something that can be altered throughout one's life. Her argument can be read as a direct rejection of the liberal value system, founded on such notions as individuality, autonomy and universal rationality. The notion of nesting multiple personalities within the same body goes against every major theory of human subjectivity, not only the Cartesian *cogito* but also Kantian or Hegelian concepts. Rosà predicts profound effects of science on the human psyche that lead her to adopt a truly transhumanist view of personality as software that can be installed, updated and removed at will from its host wetware. We should note that Rosà probably appropriated the basic idea of the scientific study of deterministic chaos from Poincaré (130–42); still, the totality of her vision, the metaphors she used and the connection she made between the evolution of science and grand social changes elevate *A Woman with Three Souls* above the status of a mere illustration.

In Rosà's novel, even though science is a powerful tool, it is not as reliable and predictable as the positivists would make it out to be. Possessed by the spectre of a female scientist from the future, Giorgina passionately speaks in a language that is completely obscure to her contemporaries. She professes a new scientific method that searches for solutions to a problem 'in the material diametrically opposed to it' (29). Giorgina's science of the future deals with matters of great complexity and thus steers clear of the determinist attitude of the classical tradition. In a sense, she anticipates the arrival of chaos theory and complexity studies which tackle physical phenomena too complex and interconnected to be approached with classical tools in hand. In yet another spark of prognostic brilliance, Rosà even employs the metaphor of a butterfly which one should study 'to put an end to social ills' (29). This example, which must have been preposterous at the time, seems far less absurd when we compare it to the famous butterfly effect – a metaphor introduced by Edward N. Lorenz in 'Predictability: Does the Flap of a Butterfly's Wings in Brazil Set off a Tornado in Texas?' As it turns out, the fate of a butterfly can, in fact, change the fate of a whole society.

Giorgina's quasi-scientific babble about butterflies 'trapped in an icebox in summer' or 'strange combinations of dyes in the laboratories of [. . .] fashion houses' actually reveals a point that is serious and crucial to understanding Futurism's attitude toward science. As with any other topic, their high praise for science actually only applied to future science, not the dominant model practised in universities, just as their nationalism was based on a purely speculative idea of a modernised Italy. The Futurist image of science is unquestionably anti-Newtonian, that is, probabilistic instead of deterministic, concerned with complex, mass phenomena instead of simple causal formulas, and, last but not least, it is performative instead of descriptive, actively engaged in transforming the world (not only explaining it). And if the scientist's role was to alter the physical environment to fit human needs, the artist's labour was to update these needs and expand the field of possible 'sensibilities'.

Immaculate Perception Does Not Exist

The Futurists had many competing ideas on the nature of the new sensibility: from Marinetti's mechanically tinged vision of humanity to the mad idea formulated by Futurist painters that sensory systems can be adjusted to become aware of vibrating waves of matter. The latter concept in particular deserves a detailed elaboration as the crowning example of an aesthetic response to the epistemological revolution brought about by the post-Newtonian, probabilistic sciences. Not only has it been meticulously elaborated in theory, primarily by Boccioni, but it was also captured in a vivid literary depiction:

> The sixteen people around you in a moving tram are in turn and at the same time one, ten, four, three; they are motionless and they change places; they are coming and going, they leap into the street, are suddenly swallowed up by a flood of sunlight, then come back and sit before you, persistent symbols of universal vibration. Or sometimes

we look at the cheek of the person with whom we were talking in the street and can see the horse which is passing at the far corner. Again: Our bodies penetrate the sofas upon which we sit, and the sofas penetrate our bodies, just as the tram rushes into the houses which it passes, and in their turn the houses throw themselves upon the tram and are merged with it. (Boccioni 2016: 175)

Though by today's standards this reads like a bizarre report on an acid trip, it is one of the few attempts by the Futurists to explain their point of view without resorting to highly abstract theoretical proclamations or literary experiments. The fragment first appeared in 'Technical Manifesto of Futurist Painting' (1910) and was included in the 'Preface to the Catalogue of the First Exhibition of Futurist Painting' two years later. In the first text, it served to illustrate how to incorporate the new physics into one's experience of the world; in the second, to conclude an argument against realism in painting and, in passing, to insult 'loud and merry imbeciles' (175) whose narrow minds are incapable of tuning their senses to the demands of modernity. In both cases, it stands out in terms of its vividness and overall clarity. Boccioni, Carrà, Russolo, Balla and Severini – the authors of the manifesto – plant a seed of doubt under the commonsensical assumption that we perceive an essentially Newtonian reality, consisting of distinct objects interacting in orderly and predictable ways. If it seems so, then certain criteria must have been met: (1) the objects in one's field of vision must have discernible edges; (2) the subject's position in relation to the objects in sight should be relatively stable. Obviously, this perspective on reality was of little interest to the Futurists, who strove to find more challenging and dizzying points of view. Most of these were highly modern, and had been unattainable just a few decades before. Finding oneself in the middle of busy traffic, watching the scenery from a speeding car, or taking a carousel ride in a chaotic amusement park[2] – all these situations overload the senses with stimuli and lead to completely new modes of perception. Objects lose their definitive shapes and blend together to form new transitory phenomena, the sense of spatial relations between things is lost, reality appears fluid and less predictable.

Futurists were among the first to recognise the historicity of perception and argue for its complete revamping. Following in the

---

[2] This was depicted by Joseph Stella, an American painter, who was inspired by the Futurists after his trip to Italy in 1910–11. He devoted three paintings to the famous funfair on Coney Island. See also: *Luna Park, Paris, 1900* by Giacomo Balla.

footsteps of Friedrich Nietzsche's doctrine of perspectivism, they recognised that 'immaculate perception' is an unscientific myth. For them, the Newtonian point of view was just one of many possibilities. For an intellectual like Newton, who lived in a secluded manor in the countryside, the world could appear static. As there were few distinct objects in his environment, their identity was easy to determine, and so were the causes of their movement. It is more feasible to fix the fall of an apple in mathematical formula when the event appears singular. In a far more complex environment rich in dynamic stimuli, such as a metropolis at the beginning of the twentieth century, one cannot rely on the same cognitive patterns. The concept of sensibility, mentioned in almost every major Futurist manifesto, relies on this understanding of perception. Sensibilities are not determined by nature – and thus no one model should be canonised and sanctified by art – but are performatively shaped by environments which change constantly in time. In other words, sensibilities are adaptive, not normative. The task of updating them lay at the heart of the Futurist aspiration to create the new Man (Milan 2009: 63).

In short, the main Futurist goal was to exert an influence on the audience and transform how people perceived the world through their novel art. This (essentially political) challenge relied on the assumption that human brains are plastic organs capable of radical transformation if exposed to new stimuli for long enough. Marinetti professed this idea, acknowledged unanimously by all the group's members, in his 'Technical Manifesto of Futurist Literature' (1912), yet another text swarming with concepts that seemed to be borrowed from the distant Future. Here he takes a strictly stochastic approach to poetry. Not only does he advise poets to distribute 'nets of images or analogies' with 'a maximum of disorder', he also grounds his proclamations in a peculiar theory of perception and communication:

> Nobody can renovate his own sensibility all at once. Dead cells are mixed together with live ones. Art is a need to destroy and disperse oneself, a great watering can of heroism that drowns the world. And don't forget: microbes are necessary for the health of the stomach and the intestines. Just so there is also a species of microbes that are necessary for the health of art – art, which is a prolongation of the forest of our arteries, prolongation which flows beyond the body and extends into the infinity of space and time. (Marinetti 2009: 124)

There are two fascinating ideas expressed by Marinetti in this short fragment, worthy of discussion in relation to the influence of science on Futurism. Firstly, the metaphor of art as microbiota, penetrating and supporting the living tissue, implies that Marinetti thought of culture as an extension of nature (biology), not a rejection or negation thereof. Possibly, Marinetti was referring to the biological theory of mutualism, introduced in 1874 by Pierre-Joseph van Beneden, which asserts that mutualistic relations exist between species, especially on a microbiological level. Yet it remains a mystery where he learned about the beneficial impact of microbiome on human health, as this view was only adopted by scientists in the late twentieth century.[3] Secondly, the biological metaphor also implies that art affects its recipients while it also infects them. The artist does not communicate with his audience on a conscious level – through symbolic exchange or rational dispute – but penetrates nervous tissues by simply exposing the senses to new stimuli. The artist plants patterns and associations in audiences' brains – symbols and 'ways of seeing' – which alter their sensibilities from the inside. Marinetti's path as an artist and entrepreneur, often sacrificing intelligibility for publicity, only confirms that he took the idea of art-as-parasite very seriously.

In his original analysis of avant-garde practices of social disruption, *Media Parasites in the Early Avant-Garde*, Arndt Niebisch considers Futurist manifestos to be 'enormous marketing campaigns' (2012: 6), intended to occupy and disrupt important channels of social communication, specifically daily newspapers like *Le Figaro*, where the 'Manifesto of Futurism' was first published. Although Niebisch does not comment on Marinetti's interests in microbiology, he concludes that the Futurists – more or less consciously – operated on the assumption that communication is primarily a viral process:

> Futurism [. . .] hijacked communication channels in order to transmit a forceful and extremely simple message to everybody. In fact, the message was supposed to be so simple that it should be 'understood' on a mere physiological level, as a mere nervous impulse, as an alarm. In this way, the Futurist aesthetics left the human being behind and conceptualized the receiver of poetry, art, and performances as an almost mechanical system whose behavior and reaction could be manipulated on a noncognitive level. (2012: 176)

Even if Cubists, surrealists, Dadaists and other avant-garde movements at the time experimented with similar forms (free verse, divisionism, collage etc.), only the Futurists, especially Marinetti, possessed

---

[3]   The only possibility, though quite improbable, is that Marinetti was somehow informed about work of Arthur Isaac Kendall, who was the first biologist to point out the importance of bacteria in the human gut and speculated about their probable function in regulating digestive processes (1909; see also Aziz 2009).

such self-awareness and cynicism. Again, in their understanding of communication primarily as manipulation (mind-infection), they brilliantly anticipated yet another important cultural revolution that would gather momentum in the decades to come, with the emergence of marketing. Like the language of advertisements, Futurist writings were not tailored for the rational subject. Like ads – in the words of Marshall McLuhan – they were 'not meant for conscious consumption. They [were] intended as subliminal pills for the subconscious in order to exercise a hypnotic spell' (McLuhan 1994: 228). Of course, all the elaborate strategies and petty tricks the Futurists used to capture the public's attention never really translated into widespread popularity. Obviously, they lacked the tools to study public opinion and, to tell the truth, they ignored the fact that people prefer to be flattered rather than insulted. Effectiveness aside, the Futurist approach to art, and to poetry in particular, stemmed from a deeply ingrained conviction that communication can be conceived in probabilistic/statistical terms. One does not need art critics looking for meanings if one can accurately measure the reach of one's text.

This intuitive receptivity was crucial, but it was only one trait of the Futurist sensibility. As I have already mentioned, there was never one model of the Futurist Man. For Marinetti at the beginning of his intellectual career, the ideal man of the times to come had to be brave, curious, well organised, disciplined, good at math ('a natural taste for numbers'), individualistic, laconic and pragmatic. In other words, he resembled a cyborg whose body and consciousness were extended by machines. He possessed an 'awareness of machines' and took advantage of the 'fusion of the instincts with the engine' (Berghaus 2009: 13). Other Futurists who also extensively discussed the topic of sensibility, like Boccioni or Benedetta Cappa, were fascinated by the ability to adapt to the dizzying complexity of the new world rather than the perfection of the factory machine. Benedetta, who joined the movement shortly after the war, came up with a completely different list of traits, among them a 'passion for depth', an aspiration to explore what is not immediately accessible to the senses, and a 'passion for difficult complexities' (2009: 280). In her eyes, the machine was not an absolute ideal for the modern human, but one of many tools for expanding the limits of cognition. Unlike her husband, Marinetti, she was not satisfied with the idea that the ultimate goal of human self-perfection was to mimic machines. She advised avoiding mechanical determination and seeking the 'thrill of unexpected attractions and repulsions'. Her Futurist mind was drawn to the improbable.

Like Benedetta, Boccioni held a nuanced point of view on the characteristics of Futurist sensibility and the importance of science in renewing human self-awareness. Although, given his strong affinity for the philosophy of Henri Bergson, Boccioni is often regarded as a eulogist of intuitionism, in reality he maintained that any attempt at developing a new conception of the human psyche should be mediated through scientific discourse. Yet his style of painting cannot be considered analytical. His main objection against cubism (. . .) and all art influenced by photography was precisely that it was analytical and motionless. However, he believed that aesthetic intuition should 'come from a world completely transformed by science' (Boccioni 1914: 68, in Milan 2009). To explain this apparent paradox I would like to suggest that Boccioni took a non-dualist point of view and assumed that emotional states are actively shaped by one's knowledge of the world. His understanding of intuition presupposed that it was not simply a non- or pre-discursive ability to grasp the external world. Intuition differs from logical reasoning, but at the same time, it is always informed by abstract or technical knowledge. Unlike many of his contemporaries at the beginning of the twentieth century, Boccioni was wary of fetishising the innocence of Nature. He wrote:

> It's a gross error to say that man has strayed from nature. It's not true! It would be like naively believing that animals are closer to nature than chemists . . . We possess a new instinct – the instinct of complexity. We grasp EVERYTHING through complexity, while those of the past gathered LITTLE through simplicity. (2016: 65)

Interestingly, Boccioni is referring not to science as such, but to the emergence of a new science – statistical mechanics – which later branched into a myriad of sub-disciplines, now called 'complexity studies'. Statistical mechanics, founded by Clark Maxwell and Ludwig Boltzmann, reinterpreted classical thermodynamics in a way that allowed them to understand energy processes, such as heat transfer or engine ignition, in probabilistic terms: as changes in the uncertainty of our knowledge about the state of a physical system. The grand significance of this theory lay in that it redefined entropy, previously understood as an irreversible loss of energy, in terms of a change in possible energetic complexions (Clarke 2002: 24–5). This was a major paradigm shift, which would later also give rise to a different understanding of what science might be. Although it was named 'statistical mechanics' (by J. Willard Gibbs in 1884, though some commentators see the discipline as less statistical than probabilistic [see Mayants 1984: 174]), it had little to do with classical mechanics. The crucial difference between this new approach and older physics was the attitude toward uncertainty. If the Newtonian scientist looked for simple formulas to cut through the complexities of empirical data and eliminate uncertainty at all cost, then their 'statistical' colleague accepted uncertainty as an irreducible component of the physical world.[4] If classical science aimed for the truth, its probabilistic offspring was content to arrive at estimations of high degrees of probability. Probably for that reason, Boccioni was able to imagine that these new non-reductionist sciences not only offered a different understanding of the universe, but could also sensitise humans to forms of order in seemingly disorderly, mass phenomena or help them perceive simple things as groupings of indefinitely smaller entities.

Boccioni conceived his paintings – in particular the *States of Mind* triptych: *The Farewells, Those Who Go, Those Who Stay* – to be direct manifestations of this new Futurist sensibility. Not only did he paint these three works twice in 1911, he also built an impressive theoretical system around them, presented in numerous manifestos, public speeches and a full book.[5] The notion of 'states of mind', though it might sound self-explanatory, is among the most intricate and convoluted in the Futurist dictionary. This was due to the fact that Boccioni based his idea on numerous assumptions that ran against the grain of common sense at the time, and was struggling to find precise words for his intuitions. In the most extensive text on this subject, the last chapter of *Futurist Painting Sculpture*, he came up with numerous definitions, most of which obscured rather than revealed his precise reasoning. He claimed that the (plastic) state of mind was simultaneously 'organization – that is, creation' (148), and 'the synthesis, or, better still, the emotional architecture, of the objects' plastic forces interpreted in their architectonic evolution' (150). He considered a painting to be an object which is created at the juncture between perception (and memory) and affective environment. He rejected both the representationalist tradition (image as representation) and epistemological dualism (the split between the objective and the subjective). A state of mind captured on canvas represented the painter's cerebral energy, excited by the environment and memories. The painter, being an energetic phenomenon, perceives energy swirling around them (light and sound waves) and combines it with electrical impulses running through the nervous system to create a work of art. The painting thus created is neither abstract nor subjective. This is why Boccioni insisted that external

---

[4] The topic of the special status of quantum mechanics has become – and still remains – the hot topic among physicists since the very advent of this field of research. See de Broglie 1953, Bohr 1985, Schrödinger 1995.

[5] In my analyses I will be referring to the second iteration of the series.

reality possesses its own psychology; for instance, reds or yellows grab the attention of living beings better than other colours. The interior (subject) and the exterior (objects) can never be truly separate from each other. As a result, the folk wisdom that beauty is only in the eye of the beholder is only partly true.

> We're therefore convinced that from the reciprocal influences of the environment with the object, from the indications of the object's plastic potential, from their force, which I have called their primordial psychology, derives the coordinating organization of the plastic state of mind, and this without the plastic force of the painting and sculpture being diminished. (2016: 146)

As I have already mentioned, Boccioni's remarks about the nature of the mind were deeply indebted to Bergson who, in *Matter and Memory*, argued adamantly against the positivist idea that perception and cognition can be reduced to elementary, neural processes inside one's brain. However, even if this assumption was considered anti-scientific by many of his peers (see Russell 1912), Bergson's argument that it is impossible to abstract experience from its physical context and represent in it immaterial formulas sounds less preposterous by today's standards.[6] In this particular case, Bergson's criticism of scientism was intended as an argument for a non-reductionist theory of the mind, conceived as an emergent entity, consisting of the whole body and its material environment. For Bergson, it would be absurd to assume otherwise, because the primary purpose of the mind is to navigate the organism through its immediate surroundings, gathering resources and avoiding threats. Bergson, and those Futurists who were inspired by his writings on perception, refuted positivist reductionism that isolated the brain from reality, not science as such.

Boccioni's images of the mind's activity occupy a weird transitional space between the retinal image, symbolic interpretation and basic pattern recognition. They are not about how the world looks or how the painter sees or imagines the world. Rather, they are documents of affection, that is, of being nervously excited by the surrounding world. Boccioni aims at capturing these intermediary moments of

---

[6] This view on cognition became more prominent only after the emergence of cybernetic psychology (see Maturana and Varela 1980). For decades after the war, they were contested by the advocates of a computational approach to the mind, but recently have been widely embraced under the 'embodied cognition' theory (see Chemero 2009).

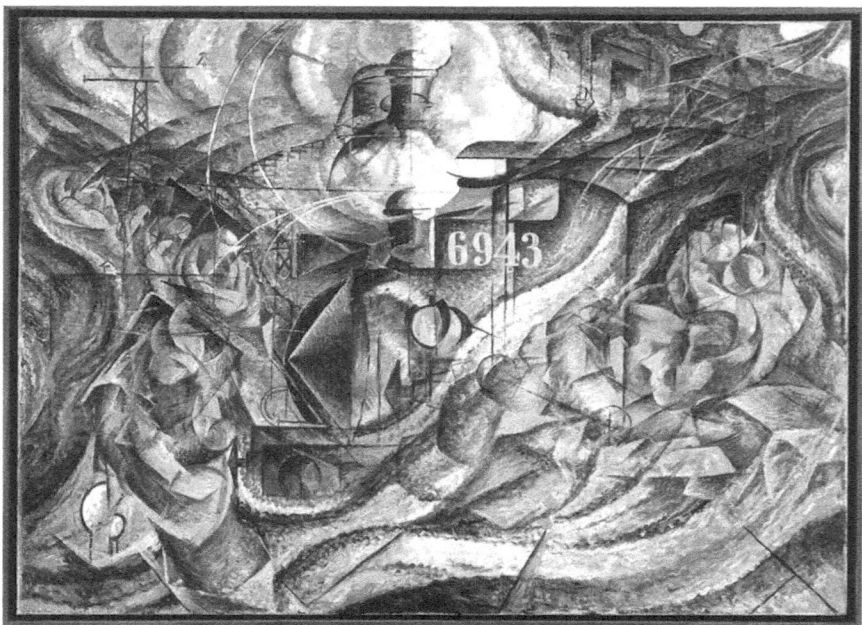

Figure 2.1 Umberto Boccioni, *States of Mind I: The Farewells* (second series), 1911, oil on canvas, DIGITAL IMAGE © 2023, The Museum of Modern Art/Scala, Florence

perception when direct sensory stimuli travels upstream through the nervous system into the visual cortex, where it is recognised. Light becomes electric impulses which then transform into primitive, pre-conscious 'afterimages', intermediate products of perception and cognition. Half-seen, half-inferred, these vague forms bear only a probabilistic relation to the external world (and consciousness). On the first picture from his famous *States of Mind* triptych, titled *The Farewells*, we see such traces of objects melted into part-abstract, part-figurative composition. Given their rectangular and sharp edges, some of these vague shapes bring to mind industrial infrastructure: railway stations, tracks, locomotives, distant factories. On top of and between those fleeting rigid forms there are ghostly silhouettes of a different kind: cylindrical and more chaotic. Probably, these are traces of people swarming on the station and smoke bursting out of the locomotive. All these forms convey only an uncertain sense of actual physical phenomena, except one – the number '6943' visible in the centre of the picture. One can only guess that this is the one detail or piece of information necessary for the painter to get his bearings, which captured the painter's attention.

The second and third painting, *Those Who Go* and *Those Who Stay*, are both more abstract and loaded with emotion. It is difficult to tell in what physical environment these states of mind occurred. In the catalogue for the 1912 exhibition of works by the Italian Futurists at London's Sackville Gallery, Boccioni described the first as follows: 'The color indicates the sensation of loneliness, anguish and dazed confusion, which is further illustrated by the faces carried away by the smoke and the violence of speed' (in Coen 1988: 121). Out of the thicket of straight lines and vague rectangular shapes, human heads emerge. In the chaotic vortex of figures and abstract shapes, one can also spot 'mangled telegraph posts' and 'fragments of the landscape through which the train has passed'. More complex shapes, primarily resembling faces, emerge out of the static of simple patterns. They are hardly images of the world as it is seen, and more representations of cerebral activity – processes that happen in a brain processing visual information. As silhouettes and faces are among the most important classes of objects processed by the visual cortex – in other words, in noisy surroundings they are the first to be recognised – they are also prioritised in Boccioni's paintings. Both pictures are less retinal and more cerebral than the first image in the triptych. They can be understood as some of the earliest artistic explorations of visual object recognition, not focusing on the optical but the cognitive dimension of seeing. For Boccioni there is no reality itself, only neural correlates of sensations awakened by the environment:

> [R]eality isn't the object but the transfiguration that the object undergoes in the process of being identified with the subject. [. . .] I'll say that an object moving quickly (a train, automobile, bicycle) appears within pure sensation as an emotive environment in the form of a sharply angled horizontal penetration, entirely different from the emotive environment in the form of perpendicular, cylindrical fullness that conveys a standing human figure. (2016: 153)

It is thus safe to say that a state of mind is not a mere subjective interpretation of a perceived phenomenon. For some reason, Boccioni distanced himself from Impressionism as well as Expressionism, traditions he knew and understood very well. First of all, a state of mind is a document of physical, that is, neuronal (nervous) excitement, aroused by an event (like boarding a train) that can be communicated to others, because, as Boccioni assumed, certain colours or shapes evoke similar emotions in most people (reds are alarming, etc.). In theory, this creates a new form of communication between the artist and their viewers; one that expresses internal states, but does not pass through the bottleneck of signification (meaning) or individual subjectivity. Being neither an impression nor an expression, it is a weird and alienating form of conveying one's psychic life, as something that is without subject. It is not by accident that all the paintings in the *States of Mind* triptych deal with some form of loss. Boccioni's mind operates as a receptor, involuntarily excited by external stimuli and the unconscious: 'the picture must be the synthesis of what one remembers and of what one sees'. Only then, after removing the ego, can the spectator 'live in the center of the picture' (Boccioni et al. 1911, in Chipp 1968: 296).

Secondly, the shadows of objects, figures or whole scenes emerging out of chaotic tangles of shapes and lines do not represent things or people in the external reality:

> [T]he composition of a plastic state of mind isn't based on the arrangement of the figure's gestures or on the expression of the eyes, faces, or poses (all of which is old literary baggage that we despise) but consists of the rhythmic distribution of forces and objects, which are dominated and directed by the very energy of the state of mind in order to compose the emotions. (2016: 149)

In other words, the painting should show how the mind is affected by reality without really representing it. For that reason, indefinite objects and figures on the canvas bear only a probabilistic relation to their sources in the external world and the memory. They can be thought of as echoes – symbols reverberating between sensory data and memories, dynamically emerging from the chaotic unconscious, only to submerge and disappear in it again a split second later. However, to better understand Boccioni's theory, we need to turn to scientific literature which could have indirectly inspired his writings.

The Futurists' Cognitive Arts

It seems that the most obvious source of inspiration for Boccioni was the philosophy of Henri Bergson, who also used the term 'states of mind' in *Matter and Memory*. However, as Maria Elena Versari noticed, Boccioni read Bergson's book after he came up with his aesthetic theory (in Boccioni 2016). She suggests that the main source of information on new developments in the sciences of the mind was not Bergson but Gaetano Previati, whose two books *The Technique of Painting* and *The Scientific Principles of Divisionism* were undoubtedly known to Boccioni prior to his theoretical efforts. Previati, an Italian painter and theoretician of Divisionism, was well informed on the latest achievements in the physiology of optics, and attempted to translate scientific knowledge into a coherent aesthetic programme. Previati believed the most important developments in the studies of perception included the idea that vision necessarily relies on both optics and psychology – one cannot reduce the act of seeing to one set of laws, physical or physiological. Versari argues that this particular idea had a lasting effect on Boccioni:

> Previati's analysis implies, however [. . .] that vision is predicated not just on retinal persistence but also on memory, which he defined as 'a purely intellectual function' that itself depends 'upon specific states of mind' [stati d'animo]. In this way, we are able to establish a link between vision and emotion. For Previati, it is through the artist's emotion (state of mind) that art distances itself from being a simple copy of reality and produces expressive forms, which are the forms most capable of provoking, in turn, emotion. (31)

Previati's concept was grounded in a theory developed in the nineteenth century by Hermann von Helmholtz in the *Treatise on Physiological Optics* (1867). This groundbreaking work gave rise to the modern science of vision by integrating the psychological, physiological and physical aspects of seeing. Crucially, his work also proved to be the first attempt to formulate a probabilistic theory of perception.

Helmholtz's path to such an innovative perspective on the human senses derived from his interest in optical illusions. He asked himself if they were mere accidental aberrations or if they revealed something essential about the nature of perception. In Volume III of his extensive treatise, he gave an example drawn from his childhood memories:

> I can recall when I was a boy going past the garrison chapel in Potsdam, where some people were standing in the belfry. I mistook them for dolls and asked my mother to reach up and get them for me, which I thought she could do. The circumstances were impressed on my memory, because it was by this mistake that I learned to understand the law of foreshortening in perspective. (1866: 283, in Gigerenzer and Murray 2015: 63)

This was one of many examples – which he complemented with experimental studies with optical equipment he developed himself – that led him to infer that visual cues from the retina are not enough to represent the world in all its depth and spatial complexity. Importantly, Helmholtz was also inspired by Kant's doctrine of transcendental idealism, which states that space and time are not directly experienced through the senses, but constitute the preconditions for experience as such. In Kant's view, the external world can be grasped only through an *a priori* framework of coordinates which exist prior to any experience. In line with this argument, Helmholtz assumed that such categories as depth or space are projected onto sensory information to construct an image of the world which we *a posteriori* mistake for truthful representations. Against common sense, he posited that the senses provide the brain with only incomplete information, from which it draws multiple unconscious inferences that organise experience within a meaningful framework. As a consequence, perception can be considered a form of gambling: the mind constantly makes new bets on what is really out there. In other words, the nature of our sensory representation of the external worlds is probabilistic – they are to an equal extent indexes and guesses. In *The Facts in Perception*, a lecture delivered in 1878, he claimed that: 'Compared to popular opinion, which in good faith assumes the complete truth of images that give us our sense of things, this residue of similarity which we acknowledge may seem very limited' (Helmholtz 1995: 347).[7]

If it is possible to view Boccioni's *States of Mind* with Helmholtz's theory in mind, it is not merely that the paintings show the psychological dimension of visual perception. Above all, they vividly depict the complexity and multidimensionality of vision as an imaginative and probabilistic activity that only in part relies on actual sensory information. Following this line of reasoning, it can be argued that the fuzzy contours of objects, faces detached from heads and other ghostly silhouettes are, in fact, probabilistic inferences – perceptual 'guesses' that the mind comes up with on the basis of limited data and its emotional state. The probabilistic dimension of Boccioni's style can also be spotted in how he paints large masses of people, especially in *The Farewells* and *Those Who Go*. In both paintings, crowds are formed from elements that can be assigned different degrees of probability – some silhouettes that catch the painter's eye resemble human beings with greater certainty, whereas others dissolve into amorphous groups where identities are lost.

Boccioni's probabilistic intuitions not only pertained to the nature of human perception and cognition. His grand aesthetic theory of 'physical

---

[7] He went on to argue that perception and reality coincide so often that we need not plunge into solipsism or other forms of epistemological pessimism. (At the end of his lecture, he stated lyrically that we should thank our senses, which miraculously gave us light and colour as responses to particular vibrations, and odour and taste from chemical stimuli. We should thank the symbols by which our senses inform us of the outside world for the spellbinding richness and the vivid freshness of the sensory world.)

transcendentalism' also included quasi-ontological reflections which, I argue, herald the arrival of the post-Newtonian world-image firmly established in the next decade by quantum physicists, in particular by Niels Bohr. Not only did the Italian painter and the Danish scientist share the same conviction that scientific progress necessitates a change of attitude toward reality, they also used a similar notion to describe it. Whereas Boccioni wrote of divisionism as 'innate complementariness' that must characterise the modern painter, Bohr coined the term 'complementarity principle' to describe the necessary abandoning of physical reductionism in the face of the intellectual challenges posed by the study of the quantum realm (1963: 1–7). For both intellectuals, the idea that one can distinguish between the objective and subjective accounts of reality was a false approximation. I will begin by summarising Bohr's view on the issue and then relate it to Boccioni's ideas.

For Bohr, complementarity did not refer to a scientific fact or physical law. His intricate philosophical system, developed on the basis of his theoretical work as a physicist, his study of modern art – cubism in particular – and his comprehensive knowledge of Western and Eastern philosophy, abolished the distinction between ontology and epistemology. In contrast to the contemporary status quo, Bohr strongly believed that science could not be divorced from its wider cultural context. Moreover, unlike his intellectual adversary, Albert Einstein, with whom he had numerous discussions on the philosophy of physics, he was able to abandon his old beliefs and adapt them to accommodate new information gathered by his colleagues and himself in laboratories. Instead of looking for ways to solve all the paradoxes of quantum physics – only to preserve the objective and deterministic image of physical reality – he proposed a new image in which some things can still make sense, but in a very different way than usual. The complementarity principle was an attempt at doing exactly so – it was 'called for to provide a frame wide enough to embrace the account of fundamental regularities of nature which cannot be comprehended within a single picture' (12). Bohr announced this idea in 1928, three years after Erwin Schrödinger presented his mathematical theory based on the hypothesis that subatomic phenomena can be perceived as waves. His theory complemented that of Heisenberg, who a few months earlier, in 1925, came up with another model, now established on the assumption that subatomic particles are, as their name suggests, in fact particles. Crucially, both theories were mathematically equivalent and produced equally accurate predictions about the behaviour of quantum systems. In other words, in the quantum realm, every entity may be described either in terms of particles, in terms of waves, or both at the same time.

For Boccioni, who died in an accident ten years before Heisenberg's and Schrödinger's work was published, the notion of complementarity meant something surprisingly parallel, though far less precise. However, if all these different meanings assigned to the notion of complementarity in *Futurist Painting Sculpture* were taken into account as building blocks of a coherent system, they amount to an aesthetic theory that could be safely compared to Bohr's onto-epistemological model. The first context in which Boccioni uses the term complementarity is ontological. He explains his point by providing the reader with a simple formula: 'Simultaneity of object + environment + atmosphere' (2016: 136). This communicates his basic belief about the very nature of reality, namely that the identity of objects is an abstraction imposed on humans by *passéist* art and science. In his view, no object ever exists other than through its relations to other objects. 'Today, our evolved mind-set no longer allows us to see an individual or an object isolated from their environments. In painting and sculpture, the essential reality of the object can only be expressed as the plastic outcome of the interaction between object and environment.' (93) It is thus entirely superficial – and intellectually outmoded – to impose identities on things or introduce discernible planes.

> What I am claiming here is not an insane abstraction, as many have thought when deriding our experiments. On the contrary, it is the static character of traditional art that is a counternatural abstraction, a violation, a separation from the real, a conception that tries to stand outside universal motion's law of unity. ([1914] 2009: 191)

In Boccioni's eyes, people really amalgamate with sofas and streetcars do enter houses. In the portrait of his mother, titled *Materia*, which in itself is a philosophical statement on the inseparability of the material world and our human understanding thereof, Boccioni most visibly puts this idea into practice. Linda Dalrymple Henderson comments that in this painting 'there is unprecedented interpenetration between object and environment' (Henderson 2002: 133). The figure in the centre of the picture is torn apart, as if exploding with dark matter. In effect, her fractured body interferes and interacts with the urban background and home interior, blurring the borders between inside and outside the represented subject. Yet this is not a form of grievance – the mother's overwhelming presence exerts a gravitational pull; reality implodes toward the centre as it explodes toward the periphery. The observer's point of view, in this case, that of Boccioni looking at his dear mother, is not hidden, but vividly exposed. The object and the observer create a complementary system of reciprocal influence: the thing cannot exist in itself, and the same rule applies to the subject, who is constituted by what he sees. This is another aspect of Boccioni's complementarism, described by yet another formula: 'Simultaneity of the internal with the external + memory + sensation' (2016: 136). The painter incorporates himself, that is, what attracts and repels him, into the image.

> The plastic potential that resides in an object is its force, that is, its primordial psychology. This power, this primordial psychology, enables us to create in our paintings new subjects which do not aim at narrative or episodic representation; instead, it coordinates the plastic values of reality, a coordination which is purely architectural and remains free of all literary and sentimental influences. (187)

As such, this is not only about the perception of the object – the thing-in-itself does not exist, it is a 'thing-towards-other-things' – psychology, paradoxically, is in the object and the subject. These are glued together and cannot be isolated from each other. 'All around us roam energies that are being observed and studied; from our bodies <u>emanate fluids of potentiality</u>, of <u>attraction or repulsion</u> (the categories of sympathy, an antipathy, and love don't interest us)' (156).

It is the painter who is torn apart by the scene and patches it together through the act of painting. At the time, Boccioni was undergoing a personal crisis, which he confessed to Severini:

> The commitment I've made is terrible, and the plastic medium appear and disappear at the moment of realization . . . Is it the chaos of judgement? What is the law? . . . Is it inside? Is it outside? Is it sensation? Is it delirium? Is it the brain? Analysis? Synthesis? . . . I don't know what! – Forms upon forms – confusion – . . . If I were to continue in this vein I could only end up killing myself. Life is certainly becoming an unbearable torment. (In Petrella 2004: 60)

Fortunately, Boccioni pulled himself out of this state which, if one takes his report literally, might be considered a psychotic break-down, and translated it into a coherent theory that expressed this sense of doubt in the firm and incontestable laws of the physical world. Obviously, his stylistic bravado and exaggerated accusations against almost every other social circle masked a genuine lack of confidence in the validity and originality of his concepts. In fact, reading about new advancements in physics gave Boccioni a sense of solace in the face of doubt and adversity. *Futurist Painting Sculpture* is filled with long passages devoted to overcoming intellectual uncertainty through heated exchanges with real and made-up adversaries. It is also clearly visible how important it was for Boccioni to ground his declarations in a socially accepted form of discourse. He even expressed this sentiment explicitly, referring to the famous statement by Einstein about the equivalence of mass and energy: 'The electrical theory of matter, according to which matter would be only energy, condensed electricity, and would exist only as force, is a hypothesis that increases the certainty of my intuition' (155).

In spite of such an evident acknowledgment of the theory of relativity, Boccioni's onto-epistemological suppositions varied significantly from Einstein's, who never came to terms with Bohr's complementarity and his subsequent argument that the mind-boggling strangeness of the quantum realm rests upon fundamental indistinguishability of the observer and the observed. Einstein's lifelong ambition was to save the objectivity of science from what he saw as erasure of scientific realism by quantum physicists. From the very beginning of his career he advocated for a strong realist position, which can be boiled down to a fairly straightforward thesis about the spatial separability of physical phenomena. Einstein claimed that it is critical for science to assume that physical phenomena can be separated from each other, because otherwise it is impossible to establish physics as an intelligible science. To put it differently, we have to assume that scientists study reality in which individual phenomena exist independently from each other and from the observer. Only then it is possible to assume that scientific observation provides us with objective information.

Boccioni and Bohr take stances against pictorial and scientific realism, respectively. According to the latter, the distinction between observed phenomena and independent observers does not apply in the subatomic realm, as the measured entity becomes entangled with the measuring device. As a result, the observed and the observing form an inseparable system, whose properties emerge at the very moment of observation. Bohr explains this paradox as follows:

> In quantum physics [. . .] evidence about atomic objects obtained by different experimental arrangements exhibits a novel kind of complementary relationship. Indeed, it must be recognized that such evidence which appears contradictory when combination into a single picture is attempted, exhausts all conceivable knowledge about the object. Far from restricting our efforts to put questions to nature in the form of experiments, the notion of complementarity simply characterizes the answers we can receive by such inquiry, whenever the interaction between the measuring instruments and the objects forms an integral part of the phenomena. (4)

By a 'single picture', Bohr means a causal explanation which describes relations between entities captured from a distance. However, subatomic processes suspend causality, because it is physically impossible to discern the basic elements in a causal relation. Quantum measurement – and, to a more or less negligible extent, every kind of measurement – alters the behaviour of the measured phenomena. For instance, to determine certain properties of a tiny photon, one has to use particles of similar proportion so that any interaction affects them reciprocally. After making a measurement, it is impossible to determine what will happen to the measured entity in the future, because it has been altered by the experiment. This makes it necessary to repeat the measurement, which again leads to the same situation, and so on. As a result, we can only make predictions about particles' behaviour without ever hoping that this essential uncertainty could be resolved, for instance, with better laboratory equipment, a rearranged experiment, or a more comprehensive theory. 'The renunciation of the ideal of causality in atomic physics which has been forced on us is founded logically only on our not being any longer in a position to speak of the autonomous behaviour of a physical object' (Bohr 1987: 87). It is not only a case of admitting that there is a limit to our knowledge about the universe. Bohr's principle goes beyond the notion of things-in-themselves developed by Kant – and held dear by Einstein – which asserts that there is a limit to knowledge about external reality, but still allows us to suppose that some form of causality exists at the ultimate level.

Bohr argues that, given the very nature of the quantum realm, every theory dealing with the behaviour of subatomic particles is inevitably probabilistic. In his view, Schrödinger's mathematical interpretation of the quantum in terms of wave should not be taken literally, but metaphorically as 'the way probabilities of our predictions would "propagate" depending on the point to which a prediction would refer' (Plotnitsky 2012: 196). 'Electron clouds' are not physical entities, like those we can observe in the sky, but abstract probability distributions. As Arkady Plotnitsky explains:

> In Bohr's ultimate view, the corresponding Schrödinger equation allows one to predict the probability of finding the electron in a given region of space at a future time, say, in one second, without describing the behaviour of the electron itself between these two experiments, one already performed and one to be performed. In other words, unlike in classical physics, once we (as 'actors') make a measurement of the position of an electron at a given point, which we can do exactly, we cannot say where the electron will exactly be at a later point. We can only estimate a probability that it will be in a certain region of space, and there is always a nonzero probability that it will not be found anywhere. (2012: 163)

Hence, it is essentially wrong, at least in Bohr's view, to assert that probabilistic prediction is only an imperfect approximation and not true knowledge, which, according to the Newtonian tradition, must be essentially causal and involve spatial separability. For Bohr, this also implies that it is practically impossible to create a visual representation of quantum processes, at least one that would succumb to the strict rules of spatio-temporal logic. Such matters must remain beyond the scope of scientific knowledge which developed in the ocularocentric paradigm (see Galison 2002: 6–7). However, this does not mean that visual aids are worthless to a quantum physicist. Bohr was profoundly interested in art, in particular in avant-garde painting. For example, in 1932, at an auction in Copenhagen, he bought *La Femme au cheval* by Jean Metzinger, a founder of cubism and its most prominent early theoretician. According to numerous biographers, prior to this event Bohr had been familiar with Europe's avant-garde movements and probably read *Du cubisme*, the first treatise on cubist painting, written by Metzinger with Albert Gleizes (1912) (see Schinckus 2017). In contrast, Einstein had a taste for a different kind of art which reflected his epistemological conservatism: he was a skilful violinist, a keen listener of Mozart, Bach and Vivaldi, and an enthusiast of classical paintings by the likes of Fra Angelico, Pierro della Francesca or Giotto. His wife, Margot Einstein, reported that 'words like cubism, abstract painting (. . .) did not mean anything to him' (in Pais 2005: 16). Of course, this striking difference in the aesthetic preferences of the two physicists does not explain subtle intricacies of their intellectual disputes. Nevertheless, it tells us something about the interconnections between seemingly unrelated realms of aesthetics and epistemology of science.

This is precisely why it is relevant to read Boccioni's aesthetic programme and interpret his art through Bohr's onto-epistemological probabilism. Although many artists in his time contemplated chance and randomness, either as a source of creativity or a metaphysical problem, his works, both literary and pictorial, were concerned with probabilistic issues in more subtle, complex and profound ways. Tormented with existential anxiety, he searched for ways to undermine commonsensical assumptions about the nature of perception and its relation to the material world. Boccioni's art, at its core, deals with the fundamental uncertainty of knowledge which does not simply describe the world, but rather takes an active part in changing it. For him, perception and intuition are performative, whereas art is not descriptive, but predictive. Both are forms of engagement in the world and thus deal with the future. Consequently, the manifesto was his preferred literary form: even stating a fact in a manifesto always serves the sole purpose of establishing a reality or predicting the future. The Futurist should never strive to capture and preserve external phenomena – or his subjective interpretations – as they are, but establish a reciprocal relation in which object and subject constitute and transform each other.

The Probabilistic Mind of the Global Art Market

The most coherent and refined example of Futurist probabilism comes from Corra and Settimelli's manifesto 'Weights, Measures, and Prices of Artistic Genius'. Not incidentally, it begins with a frontal attack on the gatekeepers of literary culture, the critics, whom they accuse of 'dogmatism in the name of nonexistent authorities' (2009: 181). They suggest replacing academic criticism, which is compromised and unfit to deal with the products of a modern sensibility, with a different profession for accessing art value – 'a measurer'. As the title plainly implies, his job would be to precisely determine the prices of works of art, so that he would have to be trained in mathematics – probability theory, statistics and combinatorics – rather than literature or philosophy. This also meant that modern art, whatever its medium and form, should not be admired in terms of its meaning, truthfulness or even beauty. Corra and Settimelli reject the notion that art calls for interpretation – along with the social institutions which profit from this demand – and also that it necessitates subjective feelings of pleasure. They declare brazenly: 'Beauty has nothing to do with art. TO DISCUSS A PAINTING OR A POEM BY STARTING WITH THE EMOTION THAT IT GIVES ONE IS LIKE TRYING TO STUDY ASTRONOMY BY CHOOSING AS ONE'S POINT OF DEPARTURE THE SHAPE OF ONE'S NAVEL' (182). In other words, art can incidentally trigger aesthetic pleasure, but given the capriciousness of such feelings, they cannot support a definition of art, or be a firm foundation for professional criticism.[8]

What is art then, and how to estimate its value? Assuming that art is just one of many manifestations of intellectual activity, Corra

---

[8] Corra and Settimelli's numerical approach to art antedates a similar theory by George D. Birkhoff, who in his 1932 book *Aesthetic Measure*, defined aesthetic value of an artwork as the ratio between its order and complexity. In a sense the Futurist idea can be considered more sophisticated as it includes the thermodynamics aspect of artistic creativity: energy consumed to establish links between distant concepts (Birkhoff 2013). It is important to note that Birkhoff's ideas lived on in the philosophy of Max Bense and early computer art, whereas the Futurists fell into oblivion. The obvious reason for that state of affairs was political, but I would also surmise that the theoretical angle of 'Weights, Measures, and Prices . . .' was irreconcilable with the algorithmic idealism of computer art.

and Settimelli argue that its products should be priced accordingly to their rarity. Although these proclamations might initially sound like textbook avant-garde provocations for their own sake, they were actually founded on a conceptual bedrock that, from a contemporary perspective, seems conceptually solid and innovative. Even if it might be easily dismissed as typical expressions of Futurist bravado, the theory developed in 'Weights, Measures, and Prices of Artistic Genius' can be also read as a unique attempt at redefining art in relation to materialist theories of the mind and the rapid onset of global capitalism. Moreover, it is unique among other Futurist manifestos in its unconditional appraisal of scientific methods of little use to the artist themselves, yet which should be embraced intelligently on an institutional level. Corra and Settimelli belonged to the group of artists gathered around the journal *L'Italia futurista*, sometimes called the 'cerebralist group' (Chessa 2012: 54–8). Together with Arnaldo Ginna, Benedetta and Russolo, among many others, they often held discussions on the role of science and its relation to art. However, in most texts, like *The Futurist Science* (1916), Corra, Settimelli and others distanced themselves from actual scientific discoveries and openly embraced the occult for its appreciation of the unknown: 'We attract the attention of all the audacious minds toward that less probed zone of our reality that comprises the phenomena of mediumism, psychism, water-divining, divination, telepathy' (in Chessa 2012: 56).

Without a doubt, despite Chessa's claims, 'Weights, Measures, and Prices of Artistic Genius' is no part of the occultist body of work written by the Futurists. It is worth mentioning in this context that Corra left his intellectual circle soon after the end of the First World War, as it gravitated more closely toward pseudo-scientific mysticism. Yet even though Corra and Settimelli make numerous claims obviously inspired by scientific theories, it is difficult to pin down their precise sources. The text's most important assertion is that measuring artistic genius rests on the premise that reasoning 'is the habit of linking ideas in a particular way' (181). The nature of this process is strictly material and involves applying 'nervous energy' to assembling elements in order. The artist is thus an intellectual worker who invests energy in making associations between ideas; they are not free to create whatever they wish but are restricted to making connections with pre-existing ideas, imposed by everyday experience. In other words, cognition follows in the footsteps of external reality and its complex system of determinations. Corra and Settimelli conclude that creativity is necessarily bound by the repeatability of everyday occurrences, and add:

If our world were different, we would reason differently: if chairs falling over typically led to deafness of the left ear in all cavalry officers, that relationship would be true for us. Thus, in every mind most notions are arranged in a definite order. For example, snow-white-cold-winter, fire-red-hot, dance-rhythm-joy. Everyone is capable of associating blue and sky. But there are other pieces of knowledge between which it is difficult to establish a relationship, because they have never been associated together, because there are no obvious similarities between them. (181)

Based on this assumption, the Futurist duo proceed to make their most important claim – art's value should be directly proportionate to the energy consumed to make surprising (unusual) associations. From their standpoint – which lies somewhere between Romantic appreciation of the exceptional and almost scientist materialism[9] – the quality of a work of art thus depends on the relative unpredictability of the links established between disparate ideas. I will elaborate on this concept in detail later, but now I would like to focus on its probable intellectual debts.

As already mentioned, it is virtually impossible to create precise footnotes for 'Weights, Measures, and Prices of Artistic Genius'. However, it is not difficult to situate it within a specific intellectual tradition, quite distant from Italy in terms of its geographical and cultural origins: namely, Scottish empiricism. Corra and Settimelli's argument is founded on the associationist principle, which states that both reasoning and perception take place by associating one mental state with the next. This idea was first given its modern shape by David Hume in his seminal *Treatise on Human Nature*, though it seems the Futurists were more inspired by his late follower Alexander Bain, who was affiliated with numerous Scottish universities throughout his long and prolific career, spanning almost the entire nineteenth century. It is possible that the Italian artists were familiar with his associationist theory of the mind, developed most comprehensively in *The Senses and the Intellect*. This impressive, over 700-page-long summary of empiricist philosophy, early neurobiology and Bain's own research was re-edited and reprinted in 1903, just in time for the Futurists to lay their hands on it. In the last section of the book, Bain even addresses the issue of artistic creativity and how it might

---

[9] For Tatarkiewicz such a mixture of materialistic epistemology and Romantic aesthetics in avant-garde practices and philosophies of art led to the emergence of Creativity as an important cultural value which took the place of more idealistic notions from the past (1980: 258).

be formulated in associationist terms (. . .). He claims that artistic labour 'consists in getting up the constituent parts from the repositories of the mind, and in choosing and rejecting until the end in view is completely answered' (642). In other words, the artist scours their mental archives to come up with 'creative associations', free of the obligations of practical reasoning. With their different social and cultural backgrounds, Bain and the Futurists drew contrasting conclusions from the same theoretical presupposition. If the former argued for the necessity of conforming the artist's imagination and creativity to social norms of intelligibility, then the latter pressed for a radically experimental method. For Corra and Settimelli, artistic genius is less about creating personal yet intelligible accounts of the world, and more about discovering unusual and complex kinships between distant ideas. As such, it is necessary to establish a professional group of expert measurers to objectively ascertain how distant and unpredictable an association really is. The virtual addressee of a work of art – a reader, a gallery-goer or a listener – is no longer necessary as the ultimate point of reference for communication, but is replaced by the futurist functionary – an artistic accountant – equipped with mathematical tables and equations, whose job would to be to provide potential buyers with probabilistic meta-analysis, much like Fisher's combined probability test, first formulated in 1925. Developed by and named after Ronald Fisher, a British statistician and geneticist, it serves to combine results from independent tests pertaining to the same general hypothesis (see Salsburg 2013). After randomising the results, the scientist is supposed to incorporate extreme probability values from each test, or 'p-values', into one test statistic. In so doing, their individual bias is removed in favour of the initial hypothesis. The Futurist procedure has a similar purpose: to remove the subjective dimension of the aesthetic experience.

Furthermore, any reference to the concepts and values of liberal, humanist culture, centred around the individual, were to be abolished in favour of scientific precision. As Corra and Settimelli demanded: 'WE MUST ELIMINATE TERMS SUCH AS SOUL, SPIRIT, ARTIST, AND ANY OTHER WORD THAT IS IRREDEEMABLY INFECTED WITH PASSÉIST SNOBBERY SUCH AS THESE; THOSE TERMS MUST BE REPLACED BY EXACT DENOMINATIONS SUCH AS: BRAIN, DISCOVERY, ENERGY, CEREBRATOR, FANTAS-TICATOR, ETC.' (186). This understanding of science and its role in modern society could not be further from the Enlightenment or positivist traditions. In fact, from this passage we can infer that the Futurist attitude toward science contradicted the basic premise of

Enlightenment thought, that the study of nature should result in its subordination to human will. In 'Weights and Measures . . .' scientific progress inevitably leads to the subordination of the human to the objectifying and alienating capacities of technology and rational discourse. If anything, science, put into practice in new technologies, diminishes human agency and sovereignty. The Futurist celebration of the mathematical mind, also praised by Marinetti in 'Geometrical and Mechanical Splendour and the Numerical Sensibility', did not stem from an actual appreciation of science as a means of bettering or rationalising society. Instead, in the eyes of the Futurists, the emerging neurosciences and new advances in mathematics paved the way for integrating art with free-market economy.[10]

In their appreciation of capitalism – conceived as a transformative force altering the very foundations of human culture – Corra and Settimelli undoubtedly drew inspiration from Georg Simmel, whose seminal work, *The Philosophy of Money* (1900), had a lasting effect on the whole Futurist movement. Although many scholars have recognised the influence of Simmel's famous essay, 'The Metropolis and Mental Life' (1903), on the formation of the Futurist aesthetic programme (see Poggi 2009: 18–20, and Whyte 2000: 364–7), I would argue that it was the aforementioned monumental study of capitalism that captured the Futurist mindset on a more basic level. The main tenet of *The Philosophy of Money* states that with the acceleration of capitalism, money becomes the primary structuring agent of culture ([1900] 2004). According to Simmel, money's rise to power could take place because it essentially works as an inhuman instrument for externalising and objectifying personal needs. As capitalism establishes a common platform for all acts of exchange, it greatly increases their scope and efficiency. However, the existing systems of values and social relations grow increasingly abstract, alienating

---

[10] It is important to note here that most Futurists shared an important biographical experience: at some point in their childhood they were all forced to migrate, for economic or personal reasons. Marinetti's family lived for a long time in Egypt, as his father, Enrico, served as a legal adviser for foreign companies engaged in the national modernisation programme. Boccioni was forced to move around the country after his parents' marriage fell apart. Carra, the son of a landowner who fell out of favour, travelled around Europe during his studies. Severini, born in the small town of Cortona, moved to Rome when he was eighteen. Luigi Russolo moved to a big city (Milan) upon completing his studies, in 1901. Rosà, born in Vienna to an aristocratic family, also left her privileged life behind, married an Italian journalist, and started an artistic career abroad. In a sense, Futurism was a way of familiarising the experience of dislocation.

and freeing individuals from their social obligations. Furthermore, Simmel states that another necessary effect of the acceleration of capitalism is cultural and epistemological relativism. Why? Because, as he argues, money and its numerous institutions do not constitute an objective (scientific) system, founded upon a metaphysical notion, such as universal utility, but rather create a dynamic environment in which values are never fixed and arise spontaneously out of complex relations between objects. In other words, value is neither dictated by objective necessity nor by purely subjective personal need. Simmel writes: 'Money becomes more and more a symbol of economic value, because economic value is nothing but the relativity of exchangeable objects. This relativity, in turn, increasingly dominates the other qualities of the objects that evolve as money, until finally these objects are nothing more than embodied relativity' (127). Money can thus be understood as a 'formula of being', by which everything 'receives meaning through each other, and have their being determined by their mutual relations' (137). Its significance rises from being a modest means of communication to the very factor that shapes the evolution of human civilisation.

This not only leads to alienation from the immediate reality of conscious needs and impulses, but even to a loss of control over the system of values. Simmel argues that any attempt at planning and taking control of capitalism, at least in its minutia, is doomed to fail, because the processes of fixing prices are largely unconscious. Hence, the capitalist market develops a mind of its own: in part conscious, and in part not. There are two simple factors that determine its operations, both beyond the scope of individual and rational control: supply (need) and demand (scarcity). The first affects what enters the market; the second determines its value. According to Simmel, scarcity is not an objective condition of goods; it depends on the difficulty of their acquisition in relation to the demand in a specific spatio-temporal context. In other words, something becomes scarce and thus valuable only conditionally when certain criteria are met. Scarcity is always relative to the demand for an object, the type of labour and the resource. Furthermore, the demand is not dictated by objective utility, like, for instance, the need for food that ensures basic sustenance. More often than not, 'the desire for the object' is completely free of biological determination. Scarcity and desire create a self-perpetuating loop. As a result, an object can be desired purely because of its scarcity – in capitalism, desire is freed from its obligation to utility. Simmel thus draws a comparison between economic and aesthetic value, understood in Kantian terms, as 'a result of increasing distance, abstraction and sublimation' (71). He sees capitalism and modern art as sharing the same essential attribute: both alienate people from first-hand experience.

Corra and Settimelli arrive at the same conclusion, though they come from the opposite direction. For them, the merger of economy with art discharges the latter of all its conventional obligations, such as arousing pleasure or conveying a sense of beauty. Artists in the age of global capitalism communicate through the market and create scarce objects for rich capitalists to invest in. Given that these objects lack any practical function, works of art can be hardly considered commodities. Their value lies in their inherent scarcity. For this reason they mustn't be confined by the old constraints, whether in terms of form, medium or topic. If the artist is committed to the market and the abstract demand to generate scarcity, then their art, as a kind of currency, needs to provide the broadest spectrum of opportunities for financial speculation. This is also why Corra and Settimelli go on to claim that the perfect functionary in the Futurist art system should be mad, as their role is none other than harnessing randomness and establishing unpredictable associations between ideas. Hence, art professionalises insanity: 'An individual who is able to construct a complicated lunacy out of his own mind has real value. A GOOD MADMAN MAY BE WORTH THOUSANDS OF FRANCS' (185). This idea goes far beyond the well-established Western tradition, dating back at least to the Romantic period, that the artist challenges social norms and the dictates of reason. It differs from this popular belief, still very strong among modernist artists eager to identify themselves as society's *enfants terribles*, in one key regard: Corra and Settimelli argue that madmen can have their place at the very core of modern society 'along with the sausage-maker and the tire manufacturer' (184). In a sense, their proposal can be read as a clever response to the exclusion of insanity from the social sphere, aptly described by Michel Foucault in *Madness and Civilization* (2001). The Futurist appraisal of insanity as a mathematically sound method of producing scarcity postulates reintroducing madmen into the very core of modern society.[11] Surprisingly, the path toward this goal leads

---

[11] A similar idea can be found in the manifesto of a Russian Futurist, Vladimir Tatlin, titled 'The Initiative Individual in the Creativity of the Collective'. According to Tatlin, the artist, called the initiative individual, 'serves as a contact between the invention and the creativity of the collective' (1919). They function as a special nod in the complex information structure of society. The mathematical code reconnects artists with society ('collective numeral'). Naturally, if Corra and Settimelli consider the possibility of measuring one's genius as a way of making profit, then Tatlin, a devoted communist, hopes that mathematics will create a universal system of social communication.

through reason and mathematical instrumentality, defining art as a form of numerical knowledge.[12]

The seemingly preposterous idea that the artist's role is to produce improbable knowledge can be assimilated even further by comparing it to a different – and socially acknowledged – method of producing value by generating randomness: mining Bitcoins. The basic mechanism behind it and other cryptocurrencies follows a logic akin to Futurist art: to create a Bitcoin, machines search through vast sets of numbers for solutions to very difficult mathematical puzzles. This is a process with low odds of success, so a great deal of trial and error is required before a valid line of code, called a 'proof of work', is generated. In other words, computers – mechanical artists – come up with random solutions to mathematical tasks until they find the rare (improbable) number. Once it is found (mined), it is used as a cryptographic key to authenticate transactions. As there is no physical point of reference for Bitcoins, the rare number *itself* is the currency. It can thus be argued that cryptocurrencies put into practice the idea that relative scarcity, regardless of material factors, equals value. However, neither generating Bitcoins nor producing improbable works of art is immaterial. In the latter case, the artist is paid for the energy expended in establishing links between remote ideas (or creative computing). In the former, the process depends on computing power which increases with every calculation. As it becomes necessary to look for new numbers in an increasingly large sphere of possibility, maintaining the cryptocurrency exhausts more and more resources. By the same token, coming up with something radically new in an oversaturated postmodern art market demands more luck/creativity than ever before.

Corra and Settimelli's ideas expressed in the manifesto turn out to be consistent with a provocative diagnosis of the contemporary art market recently penned by Hito Steyerl, a German visual artist and critical thinker who comments on the techno-socio-economic sides of the art market. According to Steyerl, 'as an alternative currency, art seems to fulfil what ether and bitcoin have hitherto only promised. Rather than money issued by a nation and administrated by central

---

[12] Almost seventy years later Andy Warhol reiterates the same idea in an even more explicit form: 'Paintings are like stocks and a dealer is like a broker. Someone makes money, then there is someone else who's really good at investing in stocks, and he tells the investor what to buy. If someone tells you to go to a good gallery rather than one that's not so good, you'll get a painting that might turn out to be worth something, a painting you like that's also a good investment. Its like having a broker tell you what stocks to buy' (in Goetzmann 1995: 25).

banks, art is a networked, decentralized, widespread system of value' (2017: 182). For Steyerl, the contemporary art world, at least in its upper echelons – distributed at art shows and auctions, or exhibited in major museums and at international biennals to be later assessed by critics and academics – now exists as a coordinated system fully integrated with global capitalism. She claims: 'There are markets, collectors, museums, publications, and the academy asynchronously registering (or mostly failing to do so) exhibitions, scandals, likes, and prices. As with cryptocurrencies, there is no central institution to guarantee value' (182). Because this system of circulating art is relatively stable yet non-centralised and beyond state control, it offers a safe haven for multimillionaires looking to invest money in times of economic crisis. In my view, Futurist tactics of rejecting traditional art institutions in favour of new social elites should be understood as an untimely attempt to bring such an ecosystem to life. Corra and Settimelli's team was composed of mensurators, fantasticators, philosophers, specialists in astronomical poetry, geniuses and madmen – wacky characters who were supposed to work side by side to create a network of production and circulation which bends to the demands of global capitalism.[13]

As such, it could be said that the notion of art as currency was not necessarily sheer provocation, just another manifestation of Futurist intellectual pomposity. On the contrary, it emerged logically from a deeper conviction that the artist was no longer obligated to create for the small community of those around them, but answered to the larger council of the (post-human) global network of networks. It is a common mistake to assume that Marinetti's nationalism glorified the nation. On the contrary, the Futurists, at least before the end of the First World War, felt the utmost contempt for Italian culture, with its backward society and antiquated politics. In Marinetti's early manifestos, all the addresses to the nation urge the necessity of rejecting its cherished ideals and transforming itself to be able

---

[13] It is worth mentioning that this affinity between speculative goals of Futurists and post-modern capitalists was cleverly spotted by Julian Rosenfeldt in his movie *Manifesto* (2015), presented in art cinemas and in museums and galleries as a multi-channel video installation. The film/installation, starring Cate Blanchett, integrates numerous artistic manifestos from different time periods – by Marinetti, Antonio Sant'Elia, Tzara and Werner Herzog, among many others – and groups them into thirteen distinct sequences, each in an everyday setting. In one of them, Blanchett, dressed as a stockbroker, recites Futurist manifestos against the background of a noisy trading floor filled with monitors and people on their phones. In this manner, Rosenfeldt presents Futurism as a movement whose closest allies were the capitalists, not Mussolini's National Fascist Party.

to compete in the global political game. In other words, Futurist art, even if labelled 'Italian', was made to circulate in the global art network and compete with the French or the Germans. Moreover, in practice, Marinetti never limited his networking activities to the small circle of Italian avant-gardists, but travelled all around Europe to establish personal and intellectual links with other movements (Russia), or create Futurist 'sleeper cells' (like one in Poland). The goal was to form an essentially international movement which only seemingly waged cultural wars in the name of the motherland. Unlike the Dadaists or cubists, Marinetti pragmatically assumed that he would need the support of the state to achieve widespread success. This understanding of Marinetti's intentions can be supported by his claims of a cultural shift in the patterns and content of human identification that had taken place at the turn of the century:

> Let me explain: men have successively conquered a sense of the house, the neighbourhood in which they live, the city, the region, the continent. Today man possesses a sense of the world; he has only a modest need to know what his forebears have done, but a burning need to know what his contemporaries are doing in every part of the globe. Whence the necessity, for the individual, of communicating with all the peoples of the earth. ([1913] 2009: 144)

Marinetti was probably among the first artists to take notice of this transition, both in human consciousness and in the very fabric of planetary affairs. I would even go so far as to argue that his ideas, expressed mainly in his early manifesto/science-fiction story titled 'Electrical War' (1911), precede similar, but far more recognised concepts like Vladimir Vernadsky's noosphere, Marshall McLuhan's global village or Paul Crutzen's Anthropocene. And although all these ideas were elaborated to sensitise the general public to different cultural processes and problems, they are all founded on the recognition of a fundamental shift in relations between *Homo sapiens* and their environment, caused primarily by the advent of new communication technologies and heavy industry. Marinetti's ideas, overlooked or consciously forgotten for political reasons, lack McLuhan's historical and theoretical depth, or Crutzen's ecological angle. However, it seems likely that he was among the first public intellectuals to become aware of the totality, multidimensionality, and possible profundity of the transformations underway in civilisation. This is the most apparent in 'Electrical War', in which Marinetti paints a fascinating picture of Earth becoming deeply interconnected with cables and wireless communication systems, thus abolishing the great divide between nature and civilisation:

Through a network of metal cables, the double force of the Mediter-
ranean and Adriatic seas climbs to the crest of the Apennines to be
concentrated in great cages of iron and crystal, mighty accumulators,
enormous nerve centers sited here and there along Italy's mountain-
ous dorsal spine. Penetrating into every muscle, artery, and nerve of
the peninsula, the energy of distant winds and the rebellions of the
sea have been transformed by man's genius into many millions of
kilowatts, spreading everywhere yet needing no wires, their fecundity
governed by the control panels, like keyboards, throbbing under the
fingers of the engineers. ([1911] 2009: 101)

The metaphors in this short passage are strikingly contemporary:
communication networks combined with geological structures make
for a planetary techno-biological superorganism (see Parikka 2015).
This image is far from modern anthropocentrism – human individuality and agency disperses and disappears in the enormous complexity of this system of interconnected systems. When Marinetti speaks
of conquering the atmosphere through control of electromagnetic
waves, he writes that 'Intelligence finally reigns everywhere' ([1911]
2009: 102). Significantly, announcing the arrival of cloud computing, he does not specify *whose* intelligence it might be – human or
otherwise.

These ideas were not incidental to Marinetti's futurology. He
kept returning to them even twenty years later, for instance, in his
late manifesto 'La Radia', written in 1933 with Pino Masnata, who
joined the movement after the war. Therein, they speculated about
the transformatory potential of electricity and wireless communication, and concluded that these technologies must eventually create
a virtual environment of an unprecedented nature. They called it
'radia', the art form of the future, and 'a pure organism of radio-
phonic sensations' which would contribute to an 'immense enlarge-
ment of space' (Marinetti and Masnata [1933] 2009: 294). Their idea
of 'radia' had little to do with the technical device they could have
known – a centralised broadcasting system critical to the existence
of the Fascist state. They spoke of a technology which never came
to be. The Futurist 'radia' bears more resemblance to the wireless
Internet, in which communication patterns emerge spontaneously
(bottom-up) and interfere with each other, than the old-fashioned
radio, which streams sound unidirectionally. As Marinetti and Masnata postulated, in the 'hybrid' reality created by wireless technology
one will not be able to differentiate between the stage and the audience. In this ecosystem communication is a 'struggle of noises'. Any
discernible sender or receiver disappears in favour of an intelligent
process which governs itself through conflict and collision.

Conclusion – Toward Probabilistic Self-awareness

'We Fight against Narcotic Symmetry' – so write Marinetti, Masnata and Marcello Puma in one of the last Futurist manifestos – 'Qualitative Imaginative Futurist Mathematics' (Marinetti, Masnata and Puma [1941] 2009). Puma, a late addition to the movement, was a student of quantum mechanics interested in 'the diffusion patterns of infectious diseases' (see Noys 2014). His presence further enriched the already obscure language of the Futurists with cryptic references to post-classical physics and mathematics. At the same time, it allowed them to finally articulate their mostly vague interest in probability. After a straightforward invocation – 'Mathematicians let us affirm the divine essence of CHANCE and RANDOMNESS' – Marinetti, Masnata and Puma go on to declare: 'Let us apply the calculus of probabilities to social life' (299). As no footnotes were attached to explain the authors' intentions, one can only speculate on the meaning of this enigmatic sentence. We can be almost certain that it should not be understood literally, as a call to use mathematics to study society. There would have been nothing subversive in this, as mathematical modelling was crucial to sociology since the very advent of the discipline. However, given that this advice is part of a long list of 'renewed scientific truths' that could be applied to reinvent common values, one can surmise that 'the calculus of probabilities' signifies a device for navigating through the complexities of modern life and looking for improbable events. Marinetti, Masnata and Puma argue strongly against the universalism of science and technology. In doing so, they oppose yet another foundational myth of Modernity, practised by post-Enlightenment utopians across the ideological spectrum. For the Futurists, science was not to be used to establish objective systems of norms and values of alleged benefit to all. On the contrary, they envisioned a science 'hostile to symmetry', one that allowed for the systematic exploration of 'the discontinuous and exceptional' (300).

Such statements contradict the common assumption about the totalitarian inclinations of the Futurists. As I have already mentioned, in the early days of the movement, and in its twilight, its members advocated a political agenda that could not have been more different than Mussolini's activity as *Il Duce*.[14] 'Qualitative Imaginative

---

[14] In an article on the history of Futurism in the 1940s, Christopher Adams criticises art historians for failing to acknowledge this fact (2013). He mentions 'Plastic Illusionism of War and Protecting the Earth', a manifesto written by Marinetti and Tullio Crali in early 1944, which praises experimental poetry – 'based in the use of pure sound and the instinctive construction of neologisms in response to

Futurist Mathematics' is one of many testaments to radically anar-
chistic inclinations of the Futurist movement and the complexity
of their views on society, which did not fit well in any major political programme of the twentieth century. Moreover, in this late text
Marinetti finally addresses the concept of probability directly, exposing its centrality for the Futurist project of the modern sensibility.
Knowingly or not, in it he expresses similar sentiments and convictions to those expounded a decade earlier by de Finetti in 'Probabilism': an inherently subjectivist aspect of probability estimations,
affirmation of individuality, relativism and rejection of absolute truth
('Scientific truth is (. . .) variable according to the individual researcher's mind' (298)). It is quite probable that Marinetti stumbled across
de Finetti's work (one can be almost certain in the case of Puma)
who, in turn, in all likelihood had been familiar with Futurist art and
philosophy. In 'Qualitative Imaginative Futurist Mathematics' the
Futurists, shortly before the end of the movement, reach self-awareness
as probabilists who not only affirm chance but also our necessarily
imperfect and fragmentary knowledge which we produce trying to
foresee the future.

external stimuli' (437) – for its politically inclusive capability. Written at the peak
of the Nazis' mass extermination of ethnic minorities, 'Futurism's proposal of
poetry that would celebrate diversity rather than division, internationalism rather
than nationalism, strikes a welcome utopian note' (438). It is worth noting that it
seems unlikely that Marinetti was moved to embrace multiculturalism by the turn
of political events. Not only did he remain critical of Hitler's politics throughout his career, he also maintained his loyalty to the Italian army fighting against
the allied forces – a hopeless move from a strategic point of view. However, in
the final months of his life, he renounced his early fascination with machines as
belonging to the 'archaeology of the avantgarde'. In his view, the Future belonged
to 'a mathematical civilization which will express itself through economical conversations gestures emotions' (in Adams 2013: 439).

Chapter 3

The Surrealists: The Probabilistic Unconscious

'Please give this a thought – a door isn't entirely real.' In this way, Jacques Lacan provoked the audience with a paradox typical of his convoluted rhetoric style during a seminar on *Psychoanalysis and Cybernetics* (1988: 301). For the contemporary reader, his statement bears an obvious resemblance to the famous quote from Lana and Lilly Wachowski's *The Matrix*. In fact, both Lacan and the little Buddhist monk Neo met at Oracle's apartment in *The Matrix* were expressing a very similar idea: the door and the famous spoon, bending sideways right in front of Neo's wondering eyes, are not entirely real, because, like any other phenomenon, they are perceived in a guessing game as 1s and 0s, as heads and tails. The door, however, is a better object for illustrating this thought. In information science, one speaks of logical gates: abstract or physical instruments that produce binary outputs, leaving the door open or closed. Neither is its correct (true) form – every static door image fails to represent the concept, as it can be in only one state or in neither of them.[1] Lacan is not concerned with computer sciences, but refers to these mathematical devices to speculate on the nature of the unconscious, in particular its computational aspect, which he calls the symbolic order – the intricate system of laws, rules and other instructions that constitute the basic architecture of the human psyche. Here, quite literally, the imaginary order of perceptual hypotheses and the real order of contingent sensory data meet at the door. The door does not open onto the world – a reality beyond the psyche – it facilitates endless loops of internal communications between expectations (beliefs) and confirmations/contradictions. In

---

[1] (The word 'door' holds an odd place in the Polish language, *plurale tantum* – it is one of few nouns that exist only in plural form; it is at the same time singular and double; the author is forever making the mistake of writing 'door are'.)

other words, the mind plays games with itself as it fruitlessly tries to grasp the external world. Making sense of the world, 'the symbol's emergence into the real begins with a wager' (192).

Although much has been written on the importance of chance (*tyche*) in Lacanian psychoanalysis (see Foster 1996, Dolar 2013), the probabilistic aspects of his theory have gone largely unnoticed, especially in the works of his most prominent commentators, like Bruce Fink or Slavoj Žižek.[2] The reason for this exclusion is simple: Lacan did what he could to cover his tracks and rarely invoked the notion of probability in his later work. However, as noted by Lydia H. Liu (2010), his engagement with cybernetics and its probabilistic epistemology may have exerted a bigger influence on his theories that he would have liked to admit. There is no doubt that Lacan's first encounters with cybernetics and game theory led him to rethink his approach to Freud and come up with his own view on the relations between the conscious and unconscious. His enthusiastic response to cybernetics neither appeared out of nowhere nor completely overturned his earlier beliefs; rather, it gave him firm ground to strike an attack on the dominance of ego in psychoanalytic theory and of rational subjectivity in positivist psychiatry. From the very beginning of his career as a psychiatrist, Lacan was drawn to theories and artistic practices that considered the human psyche in terms of (weird) machines and exposed its inherently split nature.[3] His interest in and engagement with the avant-garde, especially with the surrealists, was not incidental and was worth risking his professional reputation for, as such relations were frowned upon by his fellow professionals. They considered avant-garde art, at best, to be ridiculous and at worst to be dangerous to the public's mental health (Breton [1930] 1969: 119–23). Lacan's theory of the mind may thus serve as a point of departure for taking a closer look at artistic discourses of the avant-garde that could have potentially sensitised him to cybernetics.

In this chapter, I will be exploring the intellectual affinities between Lacan's theory, surrealism and the avant-garde concepts of unconscious, paying special attention to its probabilistic and machinic aspects. Lacan's is just one of several probabilistic theories of the mind, subjectivity and the brain that were developed around

---

[2] With the notable exception of F. Kittler (in his readings of Lacan's philosophy he considers it symptomatic of the general epistemic transitions after the world wars).
[3] This understanding of cognition became more popular with second-order cybernetics (Maturana, Varela, von Foerster, Gordon Pask, etc.).

the same time (the mid-1940s to mid-1950s) by neurophysiologists, psychiatrists and philosophers. Before I go into detail tracing Lacan's roots in the avant-garde epistemologies and aesthetics of the first half of the twentieth century, I would like to elaborate a bit more upon two similar theoretical projects: one which might have indirectly inspired Lacan, i.e. Warren McCulloch's and Walter Pitts's model of neural networks and their probabilistic epistemology, and a second developed independent of the cyberneticians, i.e. Friedrich Hayek's inquiries 'into the foundations of theoretical psychology' (1952). Presenting these three approaches together might prove valuable, inasmuch they arrive at similar conclusions while stemming from seemingly disparate intellectual traditions. McCulloch and Pitts, an experimental scientist and a mathematician, both of whom had a firm grasp of Continental philosophy, invented their model with sole reference to sources in mathematical logic, and later developed it in the context of cybernetics and information theory. Hayek, who received a degree in law but wrote mostly about the economy, rimarily advanced his theory through post-positivist philosophy. By juxtaposing these three approaches, I want to show that avant-garde art – in its constant attempts to 'tame chance' – should be considered one of the main areas of intellectual activity that has given us a probabilistic world view. Moreover, I want to argue that these three approaches supplement each other in crucial ways, presenting the idea of probabilistic thinking from different points of view: mathematical (McCulloch, Pitts), epistemological/economic (Hayek), and aesthetico-existential (Lacan).

Probabilistic Circuitry of Neural Networks

McCulloch, who wrote the seminal paper 'A Logical Calculus of Ideas Immanent in Nervous Activity' together with Walter Pitts, was, like Lacan, a psychiatrist, trained also in mathematics and professionally equipped to conduct empirical brain research. The duo's mathematical model of neurons, though grounded in formal logic, was inspired by a concrete problem in neurology: 'how to handle regenerative nervous activity in closed loops' (Schlatter and Aizawa 2008). The final text did not refer to this question, but the authors themselves considered it to be a crucial step for developing a theory of cognition that would take into account both logical reasoning and 'pathological' forms of thought (an objective that was also crucial to the surrealists). To this end, McCulloch and Pitts attempted to prove that elementary computation (with Boolean logic) could be executed on a theoretical (and vastly oversimplified) neural network. Their proof turned out to be so groundbreaking (and crucial to further research into neural networks) because it allowed for a theoretical understanding of the material act of cognition, both in real and artificial brains. McCulloch and Pitts showed it was possible to consider neuron activation in terms of logical operations. Their model assumed that neural nets were capable of computing basic propositions about the outside world which could be subsequently verified with sensory data. These propositions could later form more complex combinations and lend themselves to more intricate forms of cognition.[4] They argued that if it was possible to emulate elementary logic using nets of connected neurons acting as binary switches, 'a sufficiently large number of these simple logical devices, wired together in an appropriate manner, are capable of universal computation' (Koch and Segev 2000: 1171). Jerome Lettvin, a fellow cybernetician and a friend of McCulloch and Pitts, praised their achievement as the solution to the grand 'mind-body problem' in philosophy (1989).

McCulloch himself was more modest; from the beginning, he considered their theory to be a gross simplification of real cognitive processes, possibly a step toward transcending an outdated metaphysical dualism, but definitely not the ultimate answer. He called their model neurons 'impoverished' compared to the biological device, the brain, which 'can not only compute any Boolean function of their inputs, but many others' (1974). On the one hand, he approached the mind as a type of machine that could be taken apart and analysed with mathematical tools. In the last paragraph of 'A Logical Calculus', Pitts and McCulloch provocatively stated: 'With determination of the net, the unknowable object of knowledge, the "thing in itself", ceases to be unknowable' (McCulloch and Pitts [1943] 1990: 113). This statement was a direct attack upon psychology, as it promised to reduce the sciences of the mind to mathematically focused neurology. As rightly noted by Gualtiero Piccinini, '[i]t seems to suggest that if a subject can know the structure and past activity of her own net, then she can know things in themselves' (2020: 120). On the other hand, this early breakthrough did not lead McCulloch or Pitts to assume a straightforward positivist position. Pitts would later undertake research into probabilistic three-dimensional neural networks, which would unfortunately never see the light of day, as he burned all his notes and unfinished texts. McCulloch, in turn, became interested in

---

[4] It was pointed out by A. M. Turing in 1937 and by W. S. McCulloch and W. Pitts in 1943 that effectively constructive logics, that is, intuitionistic logics, can best be studied in terms of automata.

the epistemological consequences of their theory and, in particular, in the fundamental partiality of all knowledge, which must be bound, by nature, to the organisation of neural circuitry.

Both scholars reached a similar conclusion in another co-authored text, titled 'How We Know Universals: The Perception of Auditory and Visual Forms', published four years later, in 1947. Here, they abandoned a methodology grounded in logic and replaced it with the idea of spatially distributed analogue computation (*cybernetic sense*). Tackling the problem of image recognition in superior colliculus from a computational perspective, they touched on an even more fundamental issue, that is, the formation of invariant representations from sets of different examples. To put it simply, they wanted to know how it is possible to recognise certain objects regardless of point of view, colour or position. Devising a mathematical model for such processes would later assist the design of 'nervous nets which recognize figures in such a way as to produce the same output for every input belonging to the figure' (128), just like the ones we now routinely encounter, in facial detection software, for instance. Interestingly, in light of more recent studies in neuroanatomy, their basic intuitions about the nature of image processing proved to be correct. McCulloch and Pitts rightly assumed it is impossible to write instructions in symbolic logic for image recognition machines, because this seemingly simple activity involves computing a lot of indeterminate data. This data cannot be compressed into a fixed representation, as this would make it impossible to recognise objects in different circumstances. To solve this problem, Pitts and McCulloch proposed we consider representations in terms of 'apparitions': uncertain and fuzzy approximations of objects that simultaneously defined their specifics and could be applied regardless of context. These ghostly after-images are probabilistic approximations of a group of raw sense data, in many ways resembling the shadowy, indeterminate figures depicted in Boccioni's *States of Mind* triptych (see Chapter 2).

In 1960, McCulloch, an accomplished scientist looking back at his intellectual achievements, gave a lecture with the bewildering title 'What Is a Number, That a Man May Know It, and a Man, That He May Know a Number?' Although he did not provide the audience with a definitive answer to the title question – which could be paraphrased as follows: How it is possible for a human being to count and calculate? – he stated with utmost certainty that the solution must be formulated in probabilistic terms. He was convinced that proneness to failure was an essential quality of computation as such, especially when running on wetware, which meant that every meta-theory of knowledge should be expressed in probabilities, not certainties. Interestingly, a main source of his argument was a paper by John von Neumann, 'Probabilistic Logics and the Synthesis of Reliable Organisms from Unreliable Components',[5] which, in turn, was inspired by an earlier text by McCulloch and Pitts. Therein, von Neumann sketched out a meta-theory of computation bound by a rule of error caused by 'misbehaving neurons' or failing electronics.

> Error is viewed, therefore, not as an extraneous and misdirected or misdirecting accident, but as an essential part of the process under consideration – its importance in the synthesis of automata being fully comparable to that of the factor which is normally considered, the intended and correct logical structure. (1956: 43)

What McCulloch found so interesting and groundbreaking in this text was that von Neumann was not concerned about logic in which 'only the arguments were probable, but a logic in which the function itself was only probable' (McCulloch 1988: 9). Both agreed that it is fruitless to aim for unassailable knowledge, not simply because reality is impenetrable to our minds, but rather because every cognition is a material process bound by laws of thermodynamics. In other words, instead of devising a mathematical theory to calculate probabilities, von Neumann offered a probabilistic model for describing (and engineering) thinking machines that were real – that is, bound by the irreversibility of time and entropy.[6]

Most importantly, McCulloch and Pitt's contribution paved the way for further investigations into the subject of artificially constructed neural networks and to put their hunches into practice. Their names were mentioned alongside the most important sources in Frank Rosenblatt's paper on the 'Perceptron' (1958), the first algorithm for supervised learning of binary classifiers – mathematical functions that decide whether an input matches a specific class – ever written and executed on a real machine. However, Rosenblatt's invention was also constructive – it was built in the Cornell Aeronautical Laboratory – and descriptive. Rosenblatt was concerned with a theoretical problem that had major implications for neurobiology: How is it possible that a neural network with many random connections could function

---

[5] McCulloch refers to it as 'Toward a Probabilistic Logic'.

[6] In 'The Nature of Mental States', Hilary Putnam describes the mind as 'defined similarly to a Turing Machine, except that the transitions between "states" are allowed to be with various probabilities rather than being "deterministic" (Of course, a Turing Machine is simply a special kind of Probabilistic Automaton, one with transitional probabilities 0, 1)' (Putnam 2002: 75).

reliably? From the beginning, he was convinced that the answer should be formulated using probability theory.

> Unfortunately, the language of symbolic logic and Boolean algebra is less well suited for such investigations. The need for a suitable language for the mathematical analysis of events in systems where only the gross organization can be characterized, and the precise structure is unknown, has led the author to formulate the current model in terms of probability theory rather than symbolic logic. (387–8)

Rosenblatt assumed that every pattern recognition mechanism had to tackle the problem of randomly generated connections established during learning. For example, when a net learns to recognise cats, it may encounter some images in which a cat is partially concealed by another object that has nothing to do with the animal; for example, an art nouveau vase standing on a table. Being domesticated and exceptionally mobile, cats might appear in bizarre contexts which cannot be easily predicted. As it was impossible to avoid such randomness in data fed into the Perceptron, the device could only generate results in terms of probabilities. For Rosenblatt, this was not a matter of a temporary setback, but an essential quality of every pattern recognition machine. In a sense, this was no longer a machine in the customary sense of the word, executing specific commands with certainty. Rather, it was 'capable of a certain amount of plasticity', so that 'after a period of neural activity, the probability that a stimulus applied to one set of cells will cause a response in some other set is likely to change, due to some relatively long-lasting changes in the neurons themselves' (388).

Freedom for Probabilistic Minds!

Rosenblatt's machine was thus only programmable to a certain extent. The definition, or rather the apparition of an object was expected to emerge spontaneously in the learning process. Rosenblatt borrowed this idea of an emergent nature of perception from a relatively unknown work by Friedrich Hayek, who developed it by following an intellectual path distinct from that of the cyberneticians. Although he could have been inspired by the works of Wiener and his colleagues, whom he met in the early 1950s, his book on *The Sensory Order* was rooted in post-positivist philosophy and his lifelong intellectual struggles against the doctrine of scientism, which he considered an ideological misconception and even a grave danger to human freedom. His views on the nature of perception were intricately tied to his views on the economy which, he urged, should govern itself spontaneously, without political intervention. As early as 1920, he opposed advocates of state-governed restoration programmes like Otto Neurath, who sought to extend the wartime economy and recommended that all production should be 'centrally determined based on needs as revealed by officially collected statistics' (Caldwell 2004: 243). In contrast, Hayek believed a spontaneous order emerging from a myriad of individual decisions was superior to any system that claimed to have a scientific or objective validation. From top to bottom, Hayek's views were based on a belief in evolution as the optimal regulatory mechanism, and his observations collected in *The Sensory Order* can be interpreted as laying the groundwork for his economic evolutionism.

Hayek did not counter the prevailing economic theories by engaging his opponents on the same playing field, such as a methodological debate; rather, he devised a grand theory of cognition intended to strike a blow at the very foundations of the other doctrines. To achieve this ambitious goal, Hayek launched an attack upon the notion of immediate and pure sensory experience. This was so important because proving that this unmediated experience was impossible allowed him to show that any theory built on a scientific and rationalist image of the world could not be applied to real human beings and their behaviour. In other words, the structure of knowledge of the natural order studied by physicists, chemists, etc. was incompatible with the way our minds process information and make sense of the world. In Hayek's view, it was impossible to establish immutable laws in the sensory order because of their inherently dynamic and probabilistic nature, which prevented us from forming a static and universally applicable representation of reality. His is not the standard argument against objective knowledge, that people interpret the same phenomenal world differently. He went further, claiming that, given the distributed nature of the mind, everyone had a different phenomenal experience of the world. For him, the mind was a highly complex system of connections that constantly adapts to the environment by establishing new links between its nodes or adjusting the burden on existing ones. As data flows into this system through the senses, it is instantly interpreted by the network to confirm or reject existing models of the external world. Sensory data is captured and organised by the neural environment of the brain, which necessarily differs from individual to individual. Moreover, it is even impossible to experience, consciously or otherwise, anything apart from these connections (being formed or adjusted):

Against it we should remember that we know of no physiological mechanism which can retain anything except connexions between different events, and that, therefore, any theory of mind which is to be expressed in physiological terms must use 'experience' and 'memory' in the sense which we stress by employing the term 'linkage'. (105)

These elementary linkages make other forms of neural organisation: simple maps (semi-permanent classification apparata) and complex models. Although Hayek outlines them in detail, for our purposes it is important to mention two of their many qualities. Firstly, both maps and models, to different extents, are only 'semi-permanent', to allow for adaptation. Their stability can only be understood as 'the probability of their persistence in the face of the action upon them by the environment' (128). For example, in a very real and material sense, we cannot truly know that pink elephants do not exist, because we must be able to change our minds if we actually encounter one. And inversely, it is because our world views (models) are so deeply reinforced by prolonged exposure to certain stimuli that it is so difficult to convince someone to change their mind. Christopher Nolan's *Inception* illustrates this concept most vividly: to implant an idea in someone's head, it is necessary to entwine it in her or his system of linkages.

The second important quality of Hayek's models is their predictive function. The evolutionary success of a model – for example, choosing one route home over another – depends on its ability 'to predict the effects of different courses of action, and pre-select among the effects of alternative courses those which in the existing state of the organism are "desirable"' (124). This implies mental representations are not fixed statements about reality that can be either true or false. Rather (and so much worse for the truth), they are programs making bets on reality, whose main purpose is to 'select some elements from a complex environment as relevant for the prediction of events which are important for the persistence of the structure, and it treats them as instances of classes of events' (131). In short, the mind, as depicted by Hayek, gambles with reality to ultimately confirm itself and preserve its structure. This means no human knowledge can be considered invariable. Even the seemingly immutable laws of classical physics, preserved in precise formulas in academic textbooks, are, when executed on a human prediction machine, the brain, models like any other, used to predict outcomes and reinforce existing belief systems. Hayek employs an interesting mechanistic metaphor to support this claim. Instead of modelling the brain on the computer as a complex type of calculating machine, he compares it to 'the predictor for antiaircraft guns' (forgetting to mention Wiener, who describes the same in *Cybernetics*), a modest device which in fact was an important step in the long history of analogue computers. Just like these primitive predictors, a model in the mind 'reproduces, and experimentally tries out the possibilities offered by a given situation'. The difference between the two is not of quality but of quantity, and lies merely in the complexity of the tasks at hand.[7]

The Cybernetic Anti-Humanism of Lacan's Psychoanalysis

If Hayek wanted to bring social and economic sciences up to date with the latest discoveries in the emerging field of cybernetics, Lacan made a similar effort in the field of psychoanalysis. Though we would hardly suspect him of major engagements with information theory or early advancements in artificial intelligence when we approach his legacy through the usual lenses of literary theory and Žižek's readings, media scholars have rightly pointed out that, at its core, Lacanian psychoanalysis owes a great deal to these technical discourses.

---

[7] By arguing for a predictive interpretation of knowledge, Hayek was following in the footsteps of another Austrian philosopher, Karl Popper, who published his magnum opus, *The Logic of Scientific Discovery* (*Logik der Forschung. Zur Erkenntnistheorie der modernen Naturwissenschaft*) in 1935 (the English translation appeared fourteen years later). Contrary to most epistemologists at the time, especially those belonging to the school of logical positivism, Popper abandoned the idea that justification is a necessary requirement for scientific knowledge. Instead, he assumed that knowledge is produced in a constant process of conjectures and falsifications, of making hypotheses and verifying them experimentally. However, no theory can be conclusively proven to be an eternal truth, as it must remain no more than a heuristic tool for predicting certain behaviour. This reformulation of knowledge as essentially predictive and future-oriented allowed him to opt out of Fries's famous trilemma, which states that there are only three possible ways of justifying a scientific theory (infinite regress, dogmatism or psychologism). Popper and Hayek both acknowledged that this was, in fact, an unnecessary constraint, resulting from a flawed assumption that knowledge should represent reality as it truly is. They realised that if we abandoned such lofty expectations we could still do business as usual, coming up with new theories and designing experiments to test them. For Hayek, this was particularly important in economics, which he assigned to the complex sciences that cannot be grounded in any set of immutable laws. Both thinkers also agreed that rejecting scientific objectivism had important political and ethical significance. As I have already mentioned, Hayek developed his theory, in the first place, to strike a blow at the idea of social planning, whereas Popper concluded that a predictive and probabilistic understanding of knowledge should also be acknowledged in social and political contexts, because every system founded on a unitary conception of truth gives rise to authoritarian governance.

According to Gert Lovink, his rereading of Sigmund Freud's theory and its crucial distinction into three orders of mental functioning was fashioned after a classification system borrowed from mathematical studies of communication:

> It was Lacan who elevated psychoanalysis to the level of high-tech. His separation of the imaginary, the real and the symbolic is reflected by the trinity of storage, transmission and computing. While philosophy is still preaching 'the familiarity of one's self', psychoanalysis sticks to the view that consciousness is only the imaginary interior of medial standards. Psychoanalysis is inconceivable without cybernetics ... (Lovink 1994)

The first to notice this correspondence was Friedrich A. Kittler, who argued that Lacan's classification system actually reflected the transformation of discourses about the self ushered in by the dissemination of new media technologies: the gramophone, the typewriter and film. In his reading, despite its claims to universality, Lacan's theory described a subject that is historically specific and was fabricated by the technological environment of the twentieth century (see Kittler 1999).

Lacan's dependence on concepts and metaphors borrowed from cybernetic discourse is most apparent in the second book, containing transcripts of seminars he gave between 1954 and 1955. It is here that Lacan consolidated his theory of three orders in relation to a key notion of psychoanalysis, the ego. He provided his audience with a comprehensive discussion of cybernetics, in part to account for two concepts already present in Freud's writings but not thoroughly integrated in the grand system of psychoanalytic theory: psychic automatism and, especially, chance. From the beginning of his academic career, both notions were crucial to his original rewriting of Freud's doctrine, which he began developing in the early 1930s as he came in contact with the Parisian surrealists. As eccentric avant-garde theories were dismissed by the academic community, Lacan needed a different set of sources and references to convince a wider audience of his credibility as an original, yet serious scholar. By introducing 'thinking machines' into the conversation about the nature of the human psyche, Lacan could shift the focus from the prevailing narratives, which took the notion of ego as the true locus of cognitive abilities, to a more detached and decentralised point of view. The study of logical machines capable of solving difficult tasks without subjective comprehension, and, more importantly, conceptualising the human psyche as one of many communication systems, teaches us, he argued, that 'human language works almost by itself, seemingly to outwit us' (119). In other words, looking through the lens of cybernetic theory allows us to overthrow traditional explanations as to what constitutes thinking and how the brain performs, offering Lacan a materialistic backdrop for his novel approach to psychoanalysis, untainted by the humanist traditions dominating the academic intellectual environment. In short, he contributed to neither cybernetic research nor neurobiology, but selected certain notions and general implications of the discipline to destabilise common assumptions about the human psyche and fortify his own theory of subjectivity.[8]

A trademark concept of the Lacanian system was his assertion that the unconscious, the realm of the drives in Freudian psychoanalysis, is structured like a language. By this notion, however, Lacan understood an abstract, binary system of rules of which the conscious mind is ignorant, executed automatically. This system had more in common with computer programs capable of carrying out many different tasks, from generating sentences to editing images, than literature, which only at its most avant-garde breaks free of meaning and narrative coherence. Taking computer languages as a model for the human unconscious allowed Lacan to maintain the hypothesis of its essentially automatic and structured nature without implying that it could be investigated and interpreted as one would read literature, looking for its hidden meaning. Building on the work of American cyberneticians, Lacan argues that beyond our conscious experience is a complex realm governed by a logic specific only unto itself, and varying distinctly in every individual case, one that cannot be reduced to and expressed in a language proper to interhuman communication, structured similarly for all its users (syntax and semiotics). Communication technologies studied by cyberneticians give us irrefutable proof that such alien forms of structure as binary code exist, and our language systems can be translated into these unintelligible machinic codes. In a conversation during the twenty-second seminar, held on 15 June 1955, Lacan stated:

> [W]hen one illustrates the phenomenon of language with something as formally purified as mathematical symbols – and that is one of the reasons for putting cybernetics on the agenda – when one gives a

---

[8] For this reason, he never really mentions the writings of scientists whose work would be most relevant to his research. Lacan got acquainted with cybernetics through his friend, Georges Theodule Guilbaud, a French mathematician, who popularised American theory in France (Liu 2010), so he could have not been aware that in *Cybernetics* Wiener mentions early studies conducted by McCulloch and Pitts in neurobiology. It is, however, equiprobable that he bypassed reference to their work for the same reason he never really addressed cybernetics in his later writings or seminars – to maintain his public image of a mysterious innovator.

mathematical notation of the verbum, one demonstrates in the simplest possible way that language exists completely independently of us. . . . All this can circulate in all manner of ways in the universal machine, which is more universal than anything you could imagine. (284)

From a cybernetic standpoint there is no essential difference between human cognition and machinic computation, and, for Lacan, the possibility of translating any message into 1s and 0s proves that the prevailing perception of the nature of language as a system of symbols that pertain to extra-linguistic phenomena needing replacing by the (post-anthropocentric) universalism of binary code. Moreover, this perspective allowed him to take into account many possible relations between signifiers that do not fit into the codified rules of grammar and syntax and, more importantly, to shed some light on the material dimension of unconscious processes manifesting themselves in malfunctions, in 'jammed' communication.

During another seminar debate on 19 January 1955, this time with Jean Hyppolite in attendance, Lacan directly pointed out that he owed his views on the materiality of symbolic operations to the information sciences and to Claude Shannon's work at Bell Labs on speech encryption using statistics and probability theory, and, later, on error detection in wired communication:

> The Bell Telephone Company needed to economise, that is to say, to pass the greatest possible number of communications down one single wire. In a country as vast as the United States, it is very important to save on a few wires . . . That is where the quantification of communication started. So a start was made, as you can see, by dealing with something very far removed from what we here call speech. It had nothing to do with knowing whether what people tell each other makes any sense. . . . It is a matter of knowing what are the most economical conditions which enable one to transmit the words people recognise. No one cares about the meaning. Doesn't this underline rather well the point which I am emphasising, which one always forgets, namely that language, this language which is the instrument of speech, is something material? (Lacan 82)

Like the engineers at Bell Labs, Lacan did not intend to 'care about the meaning' of a patient's dreams or conscious introspections to successfully intervene in the libidinal economy of their unconscious. On the contrary, he advised the analyst to be suspicious of meaningful statements the patient uttered during analysis, as they might as well be attempts to mislead the therapist, to preserve their dysfunctional mental integrity. In the course of therapy, especially in its early phase, the Lacanian psychoanalyst asks the patient for samples of random speech, confessions about seemingly irrelevant thoughts that return without apparent reason, or the content of recent dreams, rather than coherent stories about life-changing events (Fink 1999: 14–15). In Lacanian psychoanalysis, manifestations of chance and meaningless automatism – the materiality of thoughts recurring with unusual frequency or intensity – play a greater role than convincingly meaningful narratives about one's motivations. For this reason, the therapist should find ways to provoke or even irritate a patient instead of establishing an honest and rational relationship. During therapy, the patient should become alert to the fact that certain manifestations of the unconscious, such as slips of the tongue or excessively repeated expressions, escape their awareness but deserve critical attention.

Lacan's interest in the materiality of psyche, manifesting itself in special attention to both singular and frequent automatisms, has obvious probabilistic overtones which could have been carried over from Shannon's model of communication. As Liu rightly points out, it was of particular significance to Lacan's theory that Shannon 'discovers a concept of the message relating to uncertainty and probability (that is, which message to choose out of x number of messages) and to the ways in which communication systems should be designed to work with the statistical pattern (which he calls "redundancy") and randomness of information (which he calls "entropy")' (314). It is often forgotten that Shannon's 'quantitative' theory of information was, in fact, a probabilistic measure: 'If the number of messages in the set is finite then this number or any monotonic function of this number can be regarded as a measure of the information produced when one message is chosen from the set, all choices being equally likely' (32). Warren Weaver, who wrote the preface to the book version of Shannon's article, expressed this using a simple mathematical example: 'If one has available say 16 alternative messages among which he is equally free to choose, then since $16 = 2^4$ so that $\log_2 16 = 4$, one says that this situation is characterized by 4 bits of information' (1949: 9–10). The logarithm in Shannon's formula was based on the number 2 and referred to the simplest possible set that served as the basis for the message model. One bit equals a binary digit. One bit is also one toss of a coin with a zero (heads) on one side and a one (tails) on the other. Two throws equal four possible results, three – eight, four – sixteen, and so on. The more choices (freedom of possibility), the more information. The more possibilities and surprises in the message, the greater its value. Mathematical information theory is thus founded on combinatorics and probability.

Shannon developed this theory in part by revisiting some old practical problems encountered by the first telegraphists in the nineteenth century. Like Shannon, they were interested in improving the efficiency of telecommunications, which were based on a system of elementary symbols almost as simple as the binary code: the Morse code (which uses three signs: a dot, a dash and a space or double space). A dot and a dash make the letter A, the letter B is a dash and three dots, etc. Shannon realised that the simplicity of Morse's language allowed for a statistical approach. This was no epochal discovery – the telegraph alphabet itself, developed in 1838, owed much of its final form to a statistical investigation of language. While assigning dots and dashes to individual letters of the English alphabet, Samuel Morse and Alfred Vail realised that it would be practical – it would save a lot of work for the telegraphists typing the messages – to assign fewer symbols to the letters that occur most frequently in English, and so they set out to determine the frequency of their occurrence. The duo visited a printing house to see which fonts were used the most and learn the actual frequencies of the letters of the alphabet. As a result, the letter E, the most often used letter in English, was assigned only one dot, while the less popular Q was codified as dash, dash, dot, dash (Burns 2004: 79–84). Shannon concluded that one could investigate each language for the frequencies of certain signs, and thus describe its order mathematically. Every letter was thus characterised not only by its form, an arrangement of lines and/or dots, but also by its relative frequency: the probability of its occurrence in a message and, more importantly, in relation to other letters in a chain. For example, in the Polish language the probability of the letter Z appearing after the letter R is relatively high, far higher than in English, whereas there are no examples of double vowels. The same applies to whole words, and even sentences. This was such an important discovery because it allowed the materiality of language to be conceptualised in terms of probabilities – an idea which, on the surface, remains rather mind-boggling – and forms of order imperceptible to standpoints of logic, meaning or even phonetics. Shannon's theory also encouraged Lacan to consider language as a system with an autonomous existence that has little to do with signification (of external reality): 'Lacan grasps this novel conceptualization of the telegraphic message and its relevance to his own work, and from this understanding he derives a notion of language that gives absolute priority to the signifier (or the letter) while banishing linguistic meaning and semantics from the sign' (Liu 314).

In an unexpected (improbable) and fascinating reading of Lacan's theory, Pedro A. Ortega, a researcher in the field of artificial intelligence, posits that Lacanian conception of subjectivity is, in many regards, analogous to the abstract mathematical model of Bayesian probability. This comparison is well founded, because Bayes's theorem formalises an approach to probability which incorporates expectations of an event's likelihood. Its specificity lies in defining probabilities as 'degrees of belief in propositions about the state of the world relative to an inquiring subject' (Ortega 2015: 1). This 'subjectivist capacity' of Bayes's theory allows us to compare it to the Lacanian model of subjectivity, which relies on various probabilistic notions. For example, the imaginary register, when interpreted probabilistically, can be explained as the locus of hypotheses about reality. Its primary function is to create assumptions (images) about itself and the surrounding world. These assumptions are given structure by the symbolic register, which can be seen as a probability space containing 'the universe of all the yes/no questions (i.e. propositions) that the subject can entertain' (3). The symbolic determines the basic structure of one's conscious and unconscious desires – what kind of relations can be established between the raw data of sensory stimuli and memory, 'making up the structure of the potential realities that the subject can hope to comprehend' (3). It is an abstract, formal system, imprinted in the nervous tissue, responsible for structuring data that goes in and out of the subject. Finally, the third register, the Real, represents 'the unintelligible, random source of external perturbations that the subject picks up and integrates into her symbolic domain in the form of sense-data, thereby setting her knowledge in motion' (2). This is not, by any means, merely the external reality. The Real as the source of randomness may appear as a chance event – an unexpected and unwanted encounter – or an internal disturbance, a glitch in the nervous system. Its inner logic is, however, completely alien and impenetrable to the symbolic and imaginary registers. The Real can only be interrogated by forming beliefs and acting upon them: 'The very notion of probability and chance presupposes the introduction of a symbol into the real' (Lacan 1988: 182).

Ortega's analogy should be taken with a large grain of salt, as it both oversimplifies Lacan's theory and makes rather liberal use of mathematical formalism. Despite these inconsistencies, the very possibility of making such a comparison (along with the fact that Ortega seems unaware of Lacan's indebtedness to cybernetics) only proves that Lacanian psychoanalysis owes more to the probabilistic epistemology of the early computer sciences than its celebrated author cared to admit when his theories began to gain publicity and devoted followers. However, as I argued at the beginning, many of the tropes Lacan picked up in cybernetics were already present in the artistic and philosophical discourses of the Parisian surrealists, with whom he was well acquainted, through the publication of some of his early works in their journals, but also through close and cordial personal relationships that famously broke up the marriage of Sylvia and Georges Bataille. The surrealist movement was, of course, decisively influenced by Freud's psychoanalysis, but diverged from his legacy in crucial respects, as well evidenced by the mutual misunderstanding between Freud and the pope of surrealism, André Breton (Davis 1973). Even if Breton attempted to gain support from Freud, their approaches to the unconscious were irreconcilable (Caws 1990: x–xix). If Freud sought to interpret dreams, the surrealists marvelled at their material concreteness. If Freud took the ego for granted, the surrealists put the paranoiac in the driver's seat of the human psyche, claiming that every form of mental self-identification is, in fact, a misidentification. Finally, unlike Freudian psychoanalysis, which excludes chance, true accidents, from mental life, the surrealists were drawn to the random adventures of the human unconscious. Not only did they feel no urge to interpret the unexpected, to turn chance into meaning, they also strove to create the perfect conditions for chance to manifest itself in its concrete, irreducible and inexplicable glory. There are many ideas to be found in surrealist prose and manifestos that bear an uncanny resemblance to Lacan's theory, such as Breton's notion of the surreality which binds the unconscious to the external world, his insistence that the unconscious should be expressed at all costs rather than repressed, or even his provocative style based on word games and puns. In my discussion of the surrealist discourse on the human psyche, I will focus only on those topics that anticipate the probabilistic conceptions of cybernetics-inspired Lacanianism. This task is not as far-fetched as it might seem, because, as Jonathan P. Eburne observes, '[i]n modelling the self-governance of mechanical and biological systems alike as automatic functions, post-war cybernetics also literalised what the surrealists had advocated in their earliest investigations: namely, that automatic functions, such as unconscious processes and chance, bore the capacity to "dictate" thought and action . . .' (2015: 63). Therefore, even if cybernetics and surrealism were founded on different discourses and developed in different social contexts, both shared a scepticism toward the epistemic relevance of positivist science and liberal culture, and, more importantly, both believed that the notion of chance had a crucial role to play in overthrowing the old epistemic paradigm.

Lacan and Dalí: Paranoia as the New Normal

Lacan's temporary alliance with the surrealists reached its peak in 1933 when fragments of his doctoral dissertation, under the titles *The Problem of Style and the Psychiatric Conception of Paranoiac Forms of Experience* and *Motives of Paranoiac Crime*, were published in two issues of *Le Minotaure*, the most prominent surrealist magazine at the time. He had encountered the group at least three years earlier, however, and was immediately captivated by their interpretations of Freud's ideas. Élisabeth Roudinesco, Lacan's biographer, argues that this encounter, especially with the young Salvador Dalí, who had just joined the group, allowed Lacan to distance himself from classical psychiatry and approach Freud's ideas from a fresh perspective (1990: 110–12), one that was inconceivable to psychiatrists and psychoanalysts alike, who were too entrenched within the axiological horizon of the middle class to call into question the categorical distinctions between the normal and the pathological.

Lacan got acquainted with Dalí's ideas through an essay, 'The Rotting Donkey', on the creative usefulness of paranoiac thought, published in July 1930 in the first issue of *Surrealisme au service de la Revolution* (Surrealism in the Service of the Revolution). Captivated less by its eccentricity and more by the similarity of Dalí's take on paranoia to his own intuitions, Lacan called the artist and arranged a meeting in his studio. Accounts of how this meeting went vary greatly. For Dalí, it was a great success and the two 'conversed for two hours in a constant dialectical tumult' (18). The artist was supposedly congratulated and praised for the scientific accuracy of his ideas. In Lacan's account, however, it was less a conversation than a monologue – Dalí perorated on and on while he was forced to sit quietly and admire the artist's genius (Roudinesco 1997: 32). Whichever version is true (although Lacan's account seems more probable, either man can be hardly trusted), the encounter pushed Lacan to reformulate his theory of paranoia and later on 'place the full anthropological significance of madness at the center of the human mind' (31), as Roudinesco reports. It is thus possible that Dalí's brief, but enlightening and provocative essay, read at just the right moment, encouraged Lacan to abandon normative distinctions between the healthy and the pathological subject, to look at paranoia and other forms of 'madness' as mere disturbances of a healthy mind and not as absolute impairments. In doing so, Lacan was able to expose that foundations of the so-called sanity are shaky and susceptible to shocks. In 'The Stinking Ass', Dalí first puts forward the claim that paranoia is not a form of cognitive exception inflicted on the unlucky few; on the contrary, every attempt at comprehending external reality is to some extent delusional. Dalí developed this idea in the following years in numerous critical essays, yet nothing proved his hypothesis better than a political event taking place at the time: Adolf Hitler's rise to power. This phenomenon captured Dalí's attention, becoming almost an obsession, much to the bewilderment or downright condemnation of his peers in the surrealist movement, who were ecstatically, yet awkwardly, sympathetic to the communist cause. Even though Dalí never embraced fascism, he remained under the spell of Hitler, and saw the Nazi dictator as a prime example of a high-functioning paranoiac who managed to impose his perverse interpretation of reality upon a whole nation. Instead of conforming to reality, Hitler succeeded in using the 'external world in order to assert its dominating idea and has the disturbing characteristic of making others accept this idea's reality' (179). By that very fact, the only difference between a 'healthy paranoiac' and a true madman lay in the social recognition and acceptance of one's delusion.

Based on this hypothesis, Dalí advanced his paranoiac-critical method – a theoretical model of cognition that harnesses the capacities of the unconscious for creative purposes. Dalí's method consisted in triggering a state of paranoia that would be then consciously ordered: 'by a paranoiac and active advance of the mind, it will be possible (simultaneously with automatism and other passive states) to systematize confusion and thus to help to discredit completely the world of reality'. This approach provided the surrealists with a fresh theoretical perspective on their experiments which, until Dalí's arrival in Paris, had mostly revolved around various methods of automatic production in literature and painting. Dalí's method differed from what the Parisian surrealists were doing at the time in that it denounced the depersonalisation of the artist in the creative act – crucial to all the practices of automatic writing or drawing that were at the centre of the movement in its early phase – in favour of conscious control over the mind's irrational impulses. For instance, in a lengthy essay on one of his most famous paintings, 'Le Mythe tragique de L'Angélus de Millet' (The Tragic Myth of Millet's *L'Angélus*), Dalí reveals his thought process and critically examines his own state of mind by discussing 'the interpretative associations securing the systematic coherence of the delirious phenomena' (1997a: 293). In this description, he attempts to achieve two things at once. First, he describes the origin of the picture as mysterious and enigmatic: 'it arose in a sudden and unthinking manner, and the emotion and confusion felt were excessive and unjustifiable for the moment by any logical explanation' (288). Second, he explains the bizarre logic behind its creation in painstaking detail – its theoretical significance and the relationship between certain symbols and events in his life or daydreams. As his own ambition is to externalise 'images of concrete irrationality' in material form (as in a painting) with utmost precision, he also expects them to be studied with scientific rigour and as objectively as possible (1997b: 258).

Despite his extravagance and his knack for provocation, Dalí, like most surrealists, considered art to be a very serious occupation. In his 'New Limits of Painting' manifesto, Dalí assumed an epistemological standpoint very similar to Lacan's, arguing it is through new hypothetical models that we arrive at a new concept of reality. And this epistemic duty of all artists obliges them to 'put their trust on probability rather than on chance' and make suppositions which 'depend more and more each day on a probability completely remote from common sense, and where truth and the absurd play a primary role' (1997c: 81). The paranoiac-critical method itself was an example of such a supposition, and for this very reason Dalí cherished the blessing given to him by the 'brilliant young psychiatrist', Lacan (Dalí 2013: 17). On the one hand, Dalí's definition, taken at face value, presents art as akin to science: unrestrained by formal protocols, but sharing the same objective, that is, the production of knowledge, changing our perception of the world, and, in effect, of the world itself. On the other, it reveals Dalí's cunning and entrepreneurship, which would later earn him the scornful pseudonym Avida Dollars, a witty anagram nickname fashioned by Breton. The latter condemned Dalí's commercial sell-out that only accelerated after the 1940s and eventually led the eccentric Spaniard to perform in a nightgown for an Alka-Seltzer commercial in 1974. Even long before that, however, by suggesting artists should calculate the probabilities of their inventions, Dalí was effectively implying they should contain chance as probability, harmonising surprise with comprehensibility instead of being simply subservient to its caprices. Thus, according to Dalí, even if the artist operates outside the confines of 'common sense', personal intuitions should still be reflected on before a conscious decision is made on what and how to paint.[9]

Dalí's intuitive grasp of probability and its epistemological significance is also evidenced by his lively interest in optical illusions, especially of double images. Buttocks that look like fruit, women becoming bearded men, or landscapes that resemble faces are common sights in his paintings, especially those that came at the beginning and the very end of his engagement with the surrealist movement (1929–39).

---

[9] A similar idea appears in André Masson's essay, bearing a significant title 'Painting Is a Wager', in which the author retrospectively criticises surrealism's long fixation on the irrational and advocates for a more heterogenous approach to painting (Masson 2001). Masson proposes to consider the creative act as a strategic activity ('serious game') which involves both 'intuition and understanding, the unconscious and the conscious' working in unison (50).

On the one hand, such imagery has long been entrenched in the history of painting, both as simple trickery for grabbing attention and as allegories endowed with deep philosophical significance, as with an Italian Renaissance painter, Giuseppe Arcimboldo, who portrayed his patrons (de)composed of fruit and other signs of *vanitas*.

In 'The Stinking Ass', Dalí ascribes an important role to double images in his paranoiac-critical method as a painterly tool for discrediting the idea of pure perception whereby the subject has direct access to an objective reality. In a sense, his double images could be regarded in terms of a popular performance of science, revealing the inner workings and complexities of human perception. Especially the early paintings or drawings, when Dalí was still formulating his theory, seem more like illustrations of a theory than artworks unto themselves. As Haim Finkelstein aptly describes these images, the objects they depicted were 'exhibiting the same degree of "probability of existence" to raise "doubts about reality"' (321). Dalí's optical puzzles 'invalidated the classical psychiatric idea of paranoia as an "error" of judgment and "reason" gone mad' (Roudinesco 1997: 31). It is thus no coincidence that in many more recent books on cognitive science, examples of double images (and related optical illusions) play a similar role, revealing flaws in common assumptions about our sensory capabilities. Above all, they offer the perfect visualisation of hidden assumptions that come into play when we recognise patterns. For instance, they hold a prominent role in Andy Clark's recent book *Surfing Uncertainty* (also discussed in Chapter 5), in which the phenomenon of double images illustrates his point about the contextual and probabilistic nature of perception. Like Dalí, Clark denounces the myth of pure perception by showing how even the simplest acts of seeing or hearing are endowed with complex networks of expectations:

> Brains like ours are constantly trying to use what they already know so as to predict the current sensory signal, using the incoming signal to select and constrain those predictions, and sometimes using prior knowledge to 'trump' certain aspects of the incoming sensory signal itself. Such trumping makes good adaptive sense, as the capacity to use what you know to outweigh some of what the incoming signal seems to be saying can be hugely beneficial when the sensory data is noisy, ambiguous, or incomplete – situations that are, in fact, pretty much the norm in daily life. (2016: 51)

Clark shows that an image or a sound is successfully perceived when a prediction matches the incoming sensory signals. Double images reveal this process, as they prevent the viewer from assigning an object to a single category. As there is no final version of the events – a horse is simultaneously a woman – the mind keeps on shifting its predictions and recognising one thing or another. Much in the spirit of the paranoiac-critical theory, Clark declares that perception is always, in fact, a 'controlled hallucination' (168).

Both the paranoiac-critical method and paintings of double images rest on a theory of the human psyche that did not exist when Dalí was crystallising his ideas. Undeniably, they had much in common with Freud's psychoanalysis and the aesthetic speculations of other members of the surrealist movement, yet they diverged from these sources in subtle yet key details. By identifying himself as a healthy paranoiac who systematically mistook reality for its subjective interpretation yet was capable of stopping his mind from running wild, Dalí drew attention to the productive and imaginative forces of the unconscious. If Freud's theory adheres to the past, aiming to establish connections between symptoms and past experiences, Dalí's reckless reinterpretation of psychoanalysis focuses on desire, which is ultimately directed toward the future. Double images, as seen and utilised by the Spaniard, capture the mind in flagranti, as it wavers in vain between two interpretations, or, to put it differently, as it hesitantly projects its expectations onto the image. As such, double images prove perception is a guessing game, actively pursuing its need to confirm its own assumptions. In turn, by practising paranoia critically, Dalí hints at a separation between the conscious self and its unconscious machinery at work in the background. As the artist probes their mind, foraging into the depths of their unconscious, they arrive at a peculiar position, inside and outside of their mental universe at the same time. Observing the marvels of his own psyche, Dalí remains sceptical and critical of its products.[10] He does not identify with melting clocks or

---

[10] One such practice, recurring in Breton's *Mad Love* on several occasions, is divination using natural phenomena formed into ambiguous patterns, the very same phenomenon that captured Dalí's attention (though in a slightly different context). He refers to the writings of Leonardo and a conversation between Hamlet and Polonius. For Leonardo, the sensory uncertainty evoked by ambiguous sights proved inspiring, an idea that also captured Breton's attention. In Shakespeare's *Hamlet*, a conversation between the young prince and Polonius about the uncanny shapes of clouds allegorises a grander theme of the drama, that is, of epistemic doubt confronted with the rigidity of the crumbling world order:

Hamlet: Do you see yonder cloud, that's almost in shape like a camel?
Polonius: By the mass and 'tis – like a camel, indeed.
Hamlet: Methinks it is like a weasel.
Polonius: It is backed like a weasel.
Hamlet: Or like a whale.
Polonius: Very like a whale.

multiple images of Lenin's head hovering over piano keys, but credits such uncanny happenings to pure chance and capitalises on these precious findings. To use Lacan's terminology, Dalí's artistic and introspective strategy consists in extracting images, associations, puns and so on from the Real of the unconscious, treated, quite literally, as a material source of randomness. In the meantime, Dalí's conscious self finds itself in a position of irreducible uncertainty in relation to his innermost self. Interestingly, he seems perfectly aware that he is conditioned to do so by the fierce competition among avant-garde artists who are confronted by the demanding task of coming up with brand new 'suppositions' that exhibit 'a probability completely remote from common sense' (1997c: 81). One can thus argue that Salvador Dalí's probabilistic self-awareness was born out of the Freudian doctrine and the avant-garde precarity, forcing the artist to instrumentalise his mind's capacities to turn them into valuable resources.

Laboratories of Randomness – Breton

Lacan's indebtedness to surrealism did not stop at Dalí's paranoiac-critical method, as he was surely familiar with artworks and texts by other members of the movement.[11] In particular, the writings of Breton – whose dominant personality united and organised the whole movement until the outbreak of the Second World War – had to provide Lacan with a rich pool of ideas from which to draw. In certain aspects, his approach to the psyche was more radical than Dalí's, who wanted to strike a compromise between his revolutionary ideas and social (financial) recognition. If Dalí wanted to tame chance as it occurs in the unconscious, then Breton was more interested in studying it as objectively as possible. However, for both artists, who were able to form and sustain a temporary strategic alliance on the basis of their mutual interests, chance became an important and highly fascinating phenomenon of the mental realm; a vivid and compelling proof that human thought and creativity cannot be reduced to conscious acts of the rational mind. There was a hidden and complex realm to be explored and brought to light, to transform the dull reality created by modern rationalists. In the preface to the reprint of the first *Manifesto of Surrealism*, Breton wrote:

> I simply believe that between my thought, such as it appears in what material people have been able to read that has my signature

---

[11] Lacan published his text ('On Criminology') in the same issue of *Minotaure* in which Breton published 'The Automatic Message'.

affixed to it, and me, which the true nature of my thought involves in something but precisely what I do not yet know, there is a world, an imperceptible world of phantasms, of hypothetical realizations of wagers lost and of lies . . . ([1929] 1969: xi)

This unconscious reality – imaginative, probabilistic and deceitful – was to be studied carefully in conditions that mimicked the procedures of the experimental sciences, as unbiased by the individual, conscious mind as possible. In this regard they resembled the experimental practices of randomisation first used in the nineteenth century (by Gustav Fechner, Wilhelm Wundt and Charles S. Peirce), widely embraced after Ronald Fisher developed his probabilistic technique for data meta-analysis.

To some extent, Breton's understanding of (and liking for) dispassionate objectivism,[12] an unusual preference for a surrealist, must have been shaped by his academic background in medical school and his actual practice during the First World War in the neurological ward of a hospital filled with severely shell-shocked victims, which left an important mark on his world view, leading him to consider surrealism a method of researching the mind that was an alternative to psychiatry and experimental psychology (Foster 1997: 1–7). Furthermore, Breton's technical preoccupation with the human psyche was influenced by French parapsychology, which drew from probability theory to randomise and objectify its speculative research. And although he never utilised probability calculus in his work, it seems he gave it special significance, that of Occam's razor, allowing him to navigate between the numerous and complicated discourses of the modern sciences, some of them functioning on the margins of academic disciplines, while maintaining a critical distance. In a lengthy footnote attached to the 'Second Manifesto of Surrealism', Breton wrote:

I think we would not be wasting our time by probing seriously into those sciences which for various reasons are today completely discredited. I am speaking of astrology, among the oldest of these sciences, metapsychics (. . .) among the modern. It is merely a question of approaching these sciences with a minimum of mistrust, and

---

[12] Looking at Breton's intellectual background, it should be noted that he probably inherited some of his appreciation of science and its revelatory values from his mentor, Paul Valéry, who fostered friendships with many renowned physicists, including Niels Bohr and Louis de Broglie, conversed with them on novel problems in science, and even visited them in laboratories (see Parkinson 2008: 47n).

for that it suffices, in both cases, to have a precise – and positive – idea of the calculus of probabilities. The only thing is, we must never under any circumstances confide to anyone else the task of making this computation in our place. ([1930] 1969: 178)

It is telling that, in this side note, Breton advises the practice of critical thought backed with the calculus of probabilities, and not some metaphysical notion like intuition, reason or common sense. He considers rational probability to be the most trustworthy tool in grounding one's knowledge about products of the unconscious. In this regard, Breton's position toward science seems both unique and exemplary of the modern epistemological crisis. On the one hand, he was a relentless materialist throughout his life, firmly rejecting supernatural explanations and metaphysical constructs (despite his interest in psychics, clairvoyance, alchemy and the occult). On the other, he was fascinated by a wide array of spontaneous and impulsive psychic phenomena that have even 'today been so grossly neglected' ([1924] 1969: 11) within the rationalist, scientific world view, like subliminal suggestion or the synchronisation of minds in crowds. To remove individual bias in reasoning, Breton placed his trust in scientific empiricism and its formal methods – once even calling automatic writing a 'veritable photography of thought' ([1921] 1996: 60) – but remained suspicious of objectivist claims about the true nature of reality. In other words, he was struggling with the question of how to marry distrust toward a middle-class, rationalist (mundane, prosaic and deterministic) representation of reality with a materialistic disregard for the supernatural. These seemingly contradictory sentiments were, in fact, not uncommon at the time, as at least since the late nineteenth century there had been a surge of interest among the scientific community in the performances of popular illusionists. For psychologists like Alfred Binet, a pioneer of psychology in France, magic became a legitimate area of study, offering an abundance of data to a determined scholar about the hidden mechanisms of human cognition, such as perceptual anticipation or the limitations of attention (Thomas, Didierjean and Nicolas 2016). Without a doubt, Breton was familiar with Binet's work, which also included studies on mental automatism, and similarly, considered magic tricks to be deliberately taking advantage of the human psyche and its non-conscious conditioning.

After the end of the First World War, Breton dropped out of medical school to pursue a career as a poet, bringing the strict experimental attitude of modern medical practices along with him. In the first 'Manifesto', he acknowledged that his fascination with automatic speech was born during the war, observing patients as they delivered semi-conscious monologues 'spoken as rapidly as possible without any intervention on the part of the critical faculties . . .' ([1924] 1969: 23). Contrary to common perception of the surrealist movement, their pope, Breton, was never keen on excess, and his rigorous attitude, combined with an authoritative personality, influenced the movement as a whole. It eventually developed into a quasi-political organisation in alliance with the Communist parties in France and the Soviet Union.[13] Even if surrealism was to wage war against positivism, this quest was motivated by the fact that rigid rationalism was seen as 'hostile to any intellectual or moral advancement' (6). Surrealist experiments with automatic writing were often meticulously prepared and preceded by weeks of reading psychological and philosophical literature. Some mimicked laboratory practices, carefully devised and executed with the utmost rigour and with complete disregard for literary effects. For example, as Mark Polizzotti describes in Breton's biography, Breton and Philippe Soupault became interested in how certain material challenges affect the content of thought, devising a special experiment to study how speed affects the content of writing:

> Breton and Soupault experimented not only with different forms but also with different writing speeds. These speeds ranged from velocity v', several times faster than 'the normal speed with which a man would undertake to recount his childhood memories'; through v ('very great') and v''', to v'' ('the greatest speed possible'). At times the images came so quickly that the two authors had to resort to abbreviations in order to get them all down. This variation of speed touched the very core of the project, for it constituted no less than an attempt to determine the limits of the bond between thought and expression. 'It had seemed to me, and still does,' Breton pointed out several years later, 'that the speed of thought is no greater than the speed of speech, and that thought does not necessarily defy language, nor even the fast-moving pen.' (1995: 96)

Breton recognised right from the start that automatic writing could serve also as a tool for the poetic imagination, and as a means for surveying the unconscious and taking control of its invisible productive powers. The control afforded by an experimental setting was crucial to him, because 'he was too wary of the psychological risks involved,

---

13 He quickly parted ways with Georges Bataille and even publicly excommunicated him in the 'Second Manifesto of Surrealism' as his interests in such transgressive subjects like incest or human sacrifice were too extreme for Breton's clinical attitude.

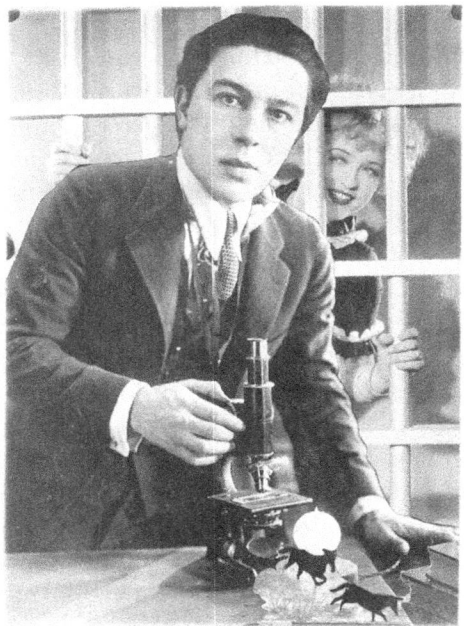

L'écriture

automatique

Figure 3.1 Breton's playful self-portrait as a scientist of the unconscious. André Breton, *Self-Portrait: Automatic Writing*, 1938, The Vera and Arturo Schwarz Collection of Dada and Surrealist Art in the Israel Museum. Copyright © ADAGP, Paris, The Israel Museum, Jerusalem

of the madness and consequent degradation that might befall too intrepid a diver into the unconscious stream' (95). Breton's passion for the improbable and unexpected paradoxically coincided with utmost methodological rigour.[14] Surrealist experiments with hypnosis, utilised as a tool to instigate automatic creativity and obtain 'the

---

[14] It is worth noting that, in contrast to many of his companions who indulged in psychedelic explorations, he refused to take any mind-altering drugs, 'possibly out of resentment against the opium that had carried off Jacques Vaché [his friend from the trenches], and quite likely out of distrust for anything that would lessen his mental self-control' (. . .). Even though he saw altering the mind and abandoning a subjective point of view to be at the heart of surrealist practices, Breton preferred to be in possession of such altered states of mind (ideally, by witnessing them) instead of being possessed by the dangerous forces of the unconscious.

desired suddenness from certain associations' (41), ended abruptly, as they became violent and unpredictable. Breton never saw falling into a hypnotic slumber as a goal in itself, it was only a means or tool for research and artistic experimentation.[15] As a trance-inducing session began, the subject themselves, or an assigned stenographer, had to note every thought uttered word for word, because, in Breton's words, 'nothing said or done is worthwhile outside obedience to that magic dictation' (1996: 91). The closest to the surrealist truth was his fellow poet, Robert Desnos,[16] who deserved such praise in Breton's eyes not for his imagination, but for his uncanny ability to shamelessly reveal his unconscious to the public: 'Desnos speaks Surrealist at will. (. . .) He reads himself like an open book, and does nothing to retain the pages, which fly away in the windy wake of his life' ([1924] 1969: 29).

More evidence of Breton's serious attitude toward artistic experiments was his undisguised affinity for formal procedures and institutions. As early as 1924, Breton, together with Antonin Artaud, established the Bureau of Surrealist Research (Bureau de recherches surréalistes), a very serious institution indeed. Among its central aims was gathering 'all the information possible related to forms that might express the unconscious activity of the mind' (Durozoi and Lecherbonnier 1972: 40). Breton announced the inauguration of the institution with a press release that abounded 'in language that [was] research and experiment-oriented', using phrases like '"experimental", "archive", "invention", "system", and "investigation"'.[17] In many ways, the surrealist institution resembled one co-founded by

---

[15] With Soupault, Breton wrote *Magnetic Fields*, his first experiment in automatic writing which produced an obscure, associative logic of the text. The relations between sentences, carrying vivid but improbable images and concepts, are always very vague. One drifts through the text which puts the sanity of an immersed mind to the test. There is no subject, no general concept, no formal discipline, nothing to secure its coherence, nothing for the logical mind to cling to. All relations between elements remain only probable as no confirmation of one's intuitions ever arrives.

[16] Desnos's imagination is so impactful not only because of its associative fervour, but also because it displays signs of surreal self-referentiality, as if the brain, not the poet, was speaking of its synaptic richness: 'We are the poet-tree pansies, the *pensées* that flower on the paths in gardens of the brain. / – Sister Anne, my St Anne, do you see nothing coming towards St Anne's? / – I see *pensées* giving their scent to words. / – We are the poet-tree words that flower on the paths of gardens of the brain, we give birth to *pensées*' (*Poesy P'Oasis* in Desnos 2017).

[17] He was not the only surrealist with a medical background. There was also Aragon, and Max Ernst, who attended lectures on psychology while a student at the university in Bonn.

Charles Richet only four years earlier, the International Institute for Metaphysics (Institut Métapsychique International), which hosted a space for organising seances with a medium, as well as a scientific laboratory devoted solely to the investigation of paranormal events (Lachapelle 2005). To convince the general public of the gravity of their undertakings, the surrealists even launched a journal, *The Surrealist Revolution* (La révolution surréaliste), in which they replicated the graphic design of scientific magazines like *La Nature* and *Science*. This deliberate modelling of their formal structure after academic and quasi-academic institutions went beyond a matter of style and self-presentation. The same austere attitude characterised surrealist experiments in the Bureau, and even earlier, in Breton's private apartment 'above Place Blanche [which] became a laboratory for the unearthing of buried voices' (Polizzotti 162).

In the early 1920s, Breton, together with Luis Aragon, Paul Éluard, René Crevel, Phillipe Soupault and others, frequently attended spiritualist spectacles. Their presence at such shows was not motivated by the need to communicate with the dead, or a belief in the accuracy of the seer's predictions. On the contrary,

> Breton took an interest in such experiments, and while he agreed to adopt the trappings of spiritualism for his own research, from the outset he drew a distinction between his goals and the spiritualists'. Breton was interested neither in communicating with the dead nor with interpreting the flow of words. Rather, it was the manifest content of the discourse itself, dredged up from the living unconscious, that attracted him. (161)

Richet's metapsychic activities were almost certainly of interest to Breton, who, in the second 'Manifesto', mentioned in passing that metapsychic researchers relied on probability theory, although he felt its usage was 'almost always out of proportion' (179). Even though many of Richet's professional activities did not come up to standards of scientific excellence, he was awarded the Nobel Prize for his work in immunology and will also be remembered as a pioneer of applying probability in psychology. Ian Hacking remarks that, even if there had been some cases of applying probability calculus to organise data in psychology:

> it remains true that, outside of astronomy and geodesy, probability had very little role in scientific inference at the time that Richet wrote. Biometrics, so valuable a source of probability ideas by 1900, and the origin of a lot of our statistical ideas, was in its infancy, or perhaps its prenatal state. Psychophysics primarily

involved modeling phenomena by probability structures and was little concerned with questions of inference. Richet, although using the most trivial of probability models for cards, was concerned above all with questions of probable inference. (1988: 438)

In 1884, Richet published an article on the transference of thought – or, in plain terms, telepathy – bearing the significant title 'Mental Suggestion and the Calculus of Probabilities'. To confirm the existence of mental transference, Richet devised a series of guessing games with cards, photographs and paintings, as well as physical activities involving spinning tables and dowsing rods (see Alvarado 2008). To objectively assess if individuals could transmit thoughts without verbal or visual cues, he analysed the results he gathered using probability calculus:

> The method that I have adopted is that of probabilities; it poses the problem thus: Given an arbitrary designation whose probability is known, does the probability of this designation change by the fact of mental suggestion? To this question our experiments allow us to reply affirmatively: For playing cards, the answer by chance should be 458, and it was 510 with suggestion on 1,833 tests. For photographs and pictures, the probable number was 42, and the acquired number was 67 on 218 tests. For experiments with the dowsing rod, the probable number was 18, and the real number was 44 on 98 tests. For experiments called spiritistic, the probable number was 3, but the real number was 17 on 124 tests. The results acquired by the calculation of serial probability are more conclusive still. (Richet 1884: 668–9, in Alvarado 2008)

Although Richet admitted that this phenomenon was uncertain and could not be explained in causal terms, he claimed to prove that the subjects' ability to guess the correct answers in his tests (playing cards, photographs and so forth) could not be attributed to pure chance. His results were never replicated and his methodology was called into question. Even Breton remained highly sceptical of its credibility as a scientific method and pointed out that these scientific experiments brought unsatisfactory results.

One can only surmise why Richet resorted to such an unpopular method. He and other researchers in the field were hard pressed to provide the scientific community with causal explanations of the phenomena they were claiming to observe. Richet speculated that there might have been an unconscious, physical process produced by the reciprocal effect of vibrating thoughts, unknowingly prophesying the advent of wireless communication technologies, which eventually made his ineptly formulated intuitions a reality. Seen from a different angle, these interests were contingent on the reality of living in the emerging mass societies, where, on an everyday basis, one could witness uncanny synchrony in herds of people. Such events had no place in the positivist world-image, which was hesitant to even consider the existence of anything that could not be causally explained (and became the subject of serious science with the advent of tracking technologies). Similarly, like many other avant-garde artists, Breton was sensitive to and intoxicated by the process of industrial modernisation and its odd manifestations in the individual and collective psyche. To some extent, Richet's fantastical vision of material reality, permeated with vibrating waves of intelligence, could also have inspired Breton's later notion of surreality, that is, reality extended and transformed by unconscious dreams and desires allowed to express themselves freely after bourgeois culture was abolished.[18] Breton did not believe in Richet's hypotheses about mental waves propagating in space and synchronising bodies at a distance, yet not unlike the latter he was attracted to these phenomena, shrouded with an uncertainty that could only be grasped rationally through probability.

It is worth noting here that Ronald Fisher illustrated his method for randomising experiments with an example that is concerned with a similarly uncertain, aesthetically almost surreal phenomenon – a lady who claimed to be able to discern if tea or milk was first poured into a cup. Fisher found her claims suspicious, because they did not make much sense from the thermodynamic point of view – the order in which liquids are poured into a receptacle should not affect the final physical form of the mixture. To expose the impostor, Fisher devised the following experiment:

> Our experiment consists in mixing eight cups of tea, four in one way and four in the other, and presenting them to the subject for judgment in a random order. The subject has been told in advance of what the test will consist, namely that she will be asked to taste eight cups, that these shall be four of each kind, and that they shall be presented to her in a random order, that is in an order not determined arbitrarily by human choice, but by the actual manipulation of the physical apparatus used in games of chance, cards, dice, roulettes, etc. (. . .) Her task is to divide the 8 cups into two sets of 4, agreeing, if possible, with the treatments received. (1935: 13–14)

---

[18] In the 1930s Breton expanded his field of interest and aimed for 'a particular philosophy of immanence according to which surreality will reside in reality itself, will be neither superior nor exterior to it (. . .), because the container shall be also the contained' ([1934] 1978: 126).

At the beginning of the experiment, Fisher assumed the lady was mistaken and formulated the 'null hypothesis' accordingly. Next, he counted the times the lady guessed correctly, and calculated combinatorically if her success could be attributed to anything other than pure chance. Unlike Richet or Breton, Fisher devised his experiment to debunk irrational claims of psychics or self-proclaimed tea-tasting experts. However, David Salsburg reports that H. Fairfield Smith, a friend of Fisher's who witnessed the experiment, disclosed that the subject actually guessed all eight cups correctly (2013: 8). Fisher omits this part of the experiment in his presentation, but if we were to take Smith's account as true, it would show that Fisher's randomisation might have been used to debunk false claims as well as to discover genuine marvels.

Another practice of probing the unconscious, both collective and individual, dear to Breton were surrealist games, an activity which, by design, involves probabilistic reasoning, both conscious (in strategies with a determinate probability of success) and unconscious (in intuitive, impulsive behaviours). However, in Breton's steady hands, games were taken seriously and played less for amusement than for artistic gain or as existential experiments: to create poems, stories or drawings in a way that circumvented subjective experience, or, as in seances, to interrogate the automatic forces of the unconscious. Some of these games were invented by the surrealists themselves, but most were simply appropriated and slightly adjusted from folk and mass culture. For example, they often played poker, both in 'dispirited' form, in which poems were used as stakes, or competitively; Breton managed to win many valuable artworks at the table (Polizzotti 1995: 345). Another popular one was Truth and Consequences (now better known as Truth or Dare), which consisted in answering questions, no matter how embarrassing or surprising they were, with complete honesty. The most prevalent game at surrealist gatherings, however, was Exquisite Corpse. It was modelled after a game called Petits Papiers, which involved multiple people writing on the same theme without any consultation and comparing the results, or drawing random words out of a hat and creating stories around them. The surrealist variant of the game required composing sentences or drawing images without seeing what previous players had written or drawn before. The name of the game was inspired by a sentence produced during one of the early rounds – 'The exquisite corpse will drink the young wine' – which Breton found particularly amusing. His preference for Exquisite Corpse over other games stemmed from the fact that it allowed for the production of objective poems, not bound to any one personal point of view, or drawings that were co-created by true chance.[19] In *Compulsive Beauty*, a seminal book on surrealism, Hal Foster suggests seeing games like Exquisite Corpse as mockeries of the dehumanising discourses and practices of the industrial age:

> Take the famous surrealist game of the *cadavre exquis* (exquisite corpse), whereby different parts of a drawing or a poem were produced by different hands oblivious to what the others had done. As is often said, such collaborations evaded the conscious control of the individual artist, but do they not also mock the rationalized order of mass production? Are these witty grotesques not also critical perversions of the assembly line – a form of automatism that parodies the world of automatization? (152)

Even if true to the surrealist spirit of anti-middle-class provocation, Foster's interpretation ignores two important things: the practical aspect of playing games within a group and Breton's personal scientific and, eventually, political ambitions. Claude Lévi-Strauss, who took part in some surrealist gatherings, recalls that even the most trivial ones, like Truth or Consequences, were taken seriously, and Breton was enraged when someone dared to open her mouth out of turn (in Polizzotti 451–2). I would prefer to think of surrealist games as practices of castrating the ego, playing a role similar to procedures of randomisation in scientific laboratories. Obviously, the surrealists playfully subverted moral norms through their satirical and erotically charged works of art, but at the same time, as explorers of the unconscious, they had to consciously relinquish their subjective control over the creation process. From this perspective, Surrealism mimics the experimental sciences in their quest for objectivity and even borrows and adapts randomisation procedures that were becoming an academic standard around the same time. Of course, their practices of objectivisation produced artefacts with an epistemic status unlike those manufactured in scientific laboratories. If the scientist was expected to produce knowledge that bore some relationship to the truth, in the form of a natural law, then the Surrealist artist was preoccupied with creating ideas that enriched the fabric of reality. Yet both professions, as Breton saw it, could benefit from the same attitude of objective detachment, aptly defined by Lorraine Daston and Peter Galison as an aspiration, prevalent in the academic discourse

---

[19] For instance, that month he participated in a surrealist game called 'Dialogue in 1928', a series of blind questions and answers, modelled on Exquisite Corpse, that were published some ten days later in *La Révolution surréaliste*.

of the nineteenth and twentieth centuries, 'to knowledge that bears no trace of the knower – knowledge unmarked by prejudice or skill, fantasy or judgment, wishing or striving. Objectivity is blind sight, seeing without inference, interpretation, or intelligence' (2007: 17). Transcending individuality is precisely the point of playing games and submitting to their rules. Caws aptly calls this perspective 'the meta-human condition' (1990: 68) – a way of looking at reality from a dissociated, non-anthropocentric point of view.

There is another argument for establishing meaningful connections between Surrealism, as defined by Breton, and the scientific practices of the twentieth century. Breton himself hints at such a possibility in a fragment of *Mad Love* ([1937] 1987), his meandering treatise on love, not as a sympathetic feeling focused upon an object, but as a predictive sensibility that orients existence toward the future and the marvellous. The essay touches on a wide array of topics, but one seems to take centre stage, returning on numerous occasions: the entanglement of internal, mental life and external reality through expectations. This entanglement of fantasy and reality manifests itself in many forms – infinite trust and loving openness or rational probabilistic calculation – which anticipate the future but also bring about real, material changes. The performative powers of prediction was a topic that had fascinated Breton for years and his view on the prophetic powers of certain individuals was, as usual, a peculiar combination of irrational attraction and rational scrutiny. In 'A Letter to Seers' of 1925 (1969), clairvoyance is both extolled and carefully observed, as during seances soothsayers reveal the inner workings of their psyches. Mediums were an object of interest for Surrealists inasmuch as they were considered to be passive instruments who registered and communicated messages from the unconscious. They belonged to the same class of subjects as the mentally ill and Surrealists in a fervour of automatic creativity – their state of mind was stripped of its critical self-awareness, the burden of bourgeois culture, becoming able to express the 'convulsively' beautiful truth of their unconscious. A seer's prediction was thus material for observation and admiration, but could also be taken as a speech act that constitutes a (sur)reality. In a sense, Breton's approach to clairvoyant language bears a resemblance to John Langshaw Austin's concept of performatives, a class of utterances that do not refer to facts (things or events that have happened) but constitute new ones instead. The most lucid example of such statements, given by Austin, is the pronouncement uttered by a priest or a clerk of marriage between husband and wife which instantly and somewhat magically transforms them as social entities (1962: 26–34). For Breton, a prophecy, when taken literally, also gives existence to realities which would not otherwise be possible. If Madame Sacco prophesised that Breton would move to China in the future, it did not actually have to happen. It sufficed that this preposterous idea was to enter Breton's mind: 'For thanks to you, I am already there' ([1925]1969: 201). According to the Surrealists, who also, as the movement evolved, were becoming increasingly materialist (in the Marxist sense of the word), mental virtualities are real, and their work might be compared to mining the depths of psyche with every automatic tool at their disposal. The role of art is to bring these two worlds even closer together, casting, as Breton declares in *Communicating Vessels*, 'a conduction wire between the far too distant worlds of waking and sleep, exterior and interior reality, reason and madness, the assurance of knowledge and of love, of life for life and the revolution, and so on' ([1934] 1990: 84). Games and other practices of automatic creativity serve as tools for this greater task of transforming reality by giving voice and agency to the unconscious realm.[20]

---

[20] The performative and predictive understanding of language, fully formulated by Breton in *Mad Love* and *Communicating Vessels* ([1934] 1990), and, to some extent, in 'The Crisis of the Object' (1936), was well integrated into the wider frame of an onto-epistemological theory of surreality. This was not by any means a fully fledged and coherent metaphysical construct. Rather, in these essays Breton tries to practise what he preaches – a style of thought in which boundaries between discourses, between the theoretically objective and personal stories of dreams and real events are breached constantly. In *Mad Love* he argues this is necessary because 'only by making evident the intimate relation linking the two terms real and imaginary I hope to break down the distinction, which seems to me less and less well founded, between the subjective and the objective. Only contemplation of this relationship leads me to wonder if the idea of causality doesn't turn out to have run quite dry' (1987: 50). Breton's attack on the cornerstone of modern epistemology, causality, was inspired by the Copenhagian philosophy of quantum physics. Taking this into account, surrealism is 'super-realism' as much as it is 'sub-realism' – a relativistic and probabilistic way of experiencing everyday reality as irreducibly entangled with our dreams and expectations (or desires) – and it is not by chance that many members of the movement, such as Roberto Matta, Wolfgang Paalen and Dalí, were eventually drawn to grapple with strangeness of the quantum realm (for a detailed historical account of the reciprocal relations between surrealism and physics, see Parkinson 2008). This sense of attraction between surrealism and science was reciprocated. For example, Paul Dirac, known for his conservative taste in art, became fond of Dalí's paintings which, in the post-war period, directly addressed quantum physics (Farmelo 2009: 347). Marie-Antoinette Tonnelat, a French theoretical physicist, also answered Breton's attempts to elicit approval from the scientific community when she recognised that: 'In physics as in painting, surrealism denies the possibility of a description which does not carry explicitly the stamp of the observer' (in Ball 2008).

Breton points out the instrumentalisation that randomness allows for:

> The whole problem of the passage from subjectivity to objectivity is implicitly resolved there, and the implications of this resolution are fuller of human interest than those of a simple technique, even if the technique were that of inspiration itself. (. . .) The novel associations of images that the poet, the artist, the scholar bring forth are comparable in that they take some grid of a particular texture, whether this texture be concretely that of a decrepit wall, of a cloud, or of anything else: some prolonged and vague sound carries this melody that we needed to hear, excluding any other. (1987: 86)

In these two regards – transcending subjectivity through chance and producing novel associations of ideas – the efforts of scientists and Surrealists, in Breton's view at least, should be perceived as analogous. Closely situating Surrealism and science is, of course, one of many examples of Breton's intellectual bravado, in his effort to raise the esteem of his movement as a serious philosophical and political project. He was, to some degree, successful. Back in the 1930s, many scientists and philosophers began wondering if the prevailing consensus on scientific knowledge and practices needed to be updated to account for the latest achievements in various areas of research, especially in quantum physics. The radical otherness (uncanniness) of some recent scientific discoveries led many academics to believe that the essence of scientific practice no longer boiled down to uncovering the hidden order of physical laws but, rather, if it should be considered in terms of creative activity, depending in equal measure on diligence and imagination. Paradoxically, improbability – of theoretical discoveries and epistemic shifts they would cause – found its place in the rational discourse of academic reasoning, formerly characterised by the certainty of logical procedures.

The Surrealist *L'Esprit de Probabilité*

This shift in understanding what constitutes scientific practice was also grounded in the latest developments in psychology, neurology and psychoanalysis. These emerging sciences made it possible to consider science as a mental activity, bound by the material conditions of the brain, and, consequently, to redefine objectivity in terms of psychological bias or concrete practices such as randomisation. Breton borrowed the idea that poets, artists and scientists were dealing with the same task, establishing 'novel associations of images', from Gustave Juvet, a Swiss mathematician and philosopher of science. He quotes a fragment from Juvet's book, *The Structure of New Physical Theories* (*Structure des nouvelles théories physiques* [1933]):

> The surprise created by a new idea or association of ideas is surely the most important element in the progress of the physical sciences, for it is astonishment that excites logic, which is always rather cold, and that forces scientists to make new connections. But the ultimate cause of progress, the reason for our surprise itself, has to be sought in the force fields that new associations of ideas set up in our minds, fields whose strength measures the good fortune of the scientists lucky enough to bring those ideas together. (173–4)

Breton's attention was clearly captured by Juvet's notion of psychic 'force fields': a materialist yet highly elusive concept, and a perfect fit for his trademark philosophical mixture of unemotional, analytic rationality and his penchant for occultism and other cultural forms embracing epistemic ambiguity, describing the electric loci of creative production. The same fragment appeared in Gaston Bachelard's *The New Scientific Spirit*, published three years prior to Breton's essay, in which the former diagnosed a crisis in modern science and the gradual but inevitable transformation of its epistemological grounds beyond the Cartesian paradigm. This was no coincidence, as Breton and Bachelard were both aware of each other's work. In 1936, Bachelard wrote a short essay with a telling title, 'Surrationalism', in which he postulated that science must benefit from randomness – 'to allow human reason its function of turbulence and aggressiveness' (2016: 78) – and to this end it had to develop new methods of reasoning, mimicking Surrealist practices of the experimental investigation of dreams. In turn, Breton mentioned Bachelard's name in 'The Crisis of the Object', a manifesto wholly based on a chapter of *The New Scientific Spirit*. Moreover, as Gavin Parkinson mentions, in Bachelard Breton 'discovered a learned figure in sympathy with some of the aims of Surrealism, whose academic status promised to strengthen its standing and its claim to historical authority in circles beyond those of poetry and art' (Parkinson 2008: 69). Caws even concludes that both the philosopher and the artist, through intellectual dialogue that eventually lasted until the 1960s, centred their grand projects – 'phenomenology of poetic imagination' and surrealism respectively – on the same fundamental beliefs 'in the union of the real and the unreal, same optimism based on the openness of the mind to all possibilities, on the power of the poet to transform the universe' (1964: 310). To put it in simple terms, they agreed that a true 'sense of wonder' – accompanied by 'lucid vision' – should prevail in science and art for true transformation of humanity to take place.

In turn, Bachelard turned to Juvet to support his utopian concept of 'open-minded rationalism', an attitude of 'genuine surprise at the implications of theoretical speculation' (1984: 173). He argued that this approach is required to navigate the unfamiliar terrain of modern science which, at its core, is no longer rooted in the early-modern rationalism of Descartes, Bacon and Newton. The methods and epistemic foundations of science had changed dramatically, argued Bachelard, and apart from various paradigmatic shifts like quantum physics or non-Euclidean geometry, the orientation of scientific practice had shifted since the beginning of the twentieth century toward a creative synthesis that was essentially at odds with Cartesian reductionism. Unlike the old science, founded on the belief in 'simple natures' gradually uncovered in a linear process, the post-Cartesian scientist dealt with theoretical relations, irreducibly complex and open to ever-new transformations. On this new epistemic horizon 'a subtle transition from a realist to a probabilistic view' (80) took place, gradually substituting certainty with likelihood as the measure of scientific accuracy. This change was most apparent in physics, where mathematical formalisations of quantum mechanics represented only probabilities of certain events happening in certain predefined conditions. Insisting on 'ultraprecise definitions' of elementary notions thus became pointless, 'leaving behind nothing more substantial than a card player's hope of winning a hand of poker' (83). Even chemistry, long considered the most 'substantial' of all sciences, when confronted by the large numbers of data collected during experiments, 'had the substance slowly drained out' (84). At this point in the history of science, rationalism, understood as a method of inductive inquiry, took precedence over, or even contradicted, the doctrine of realism, which posited the reality of things-in-themselves, of determinate objects and determinate relations between them. Breton sided with Bachelard when he said that poetry – as an instrument of a free imagination – should side with post-Newtonian science to undermine a commonsensical approach to reality that can only experience a world made of concrete, finite objects ([1936] 1972).

The Bachelardian view of science as an enterprise of embodied, innovation-driven practices did not capture much attention in the scientific community at the time and the book was not translated into English until 1984. However, in its content and general assumptions about the essence of science, the book anticipated a fundamental shift in the Anglo-Saxon philosophy of science that occurred after the Second World War, under the influence of cybernetics and Popperian post-positivism. Daring to breach the line dividing the arts and the sciences, Bachelard was able to (fore)see many trends within both fields that only became apparent and widely accepted in the decades to come. This divide, existing in practice to this very day, was famously described by the British physico-chemist and writer C. P. Snow, who coined the notion of 'two cultures' (1959): of those savvy in the Second Law of Thermodynamics and of those fluent in Shakespeare. And although Snow's view was critical of the divide, it was essentially an iteration of a long-held belief in European culture that *l'esprit de finesse*, the intuitive perception of the world, and *l'esprit de géométrie*, the mathematical mind, were fundamentally incompatible. This opposition, noted and commented on by Blaise Pascal, was explained by postulating an essential and irreducible difference between a direct and a rational world view. With the success of probability and the emergence of *l'esprit de probabilité*, which ironically originated in Pascal himself, this view was already outdated in principle, because art had begun appropriating methods of randomisation and objectivisation, while, at the same time, science, transformed by probabilistic discourses and unbound from a Newtonian world view, had become the avant-garde art itself, at least in the sense of being consumed by the urge to produce new and disarming results.

Chapter 4

Duchamp and Musil: Two Probabilistic Eccentrics, or Men without Qualities

In this chapter, I will mostly speak about the role of probabilistic concepts in Marcel Duchamp's highly abstract ways of making art. There is much evidence of Duchamp's interest in the notion of probability: enigmatic comments scattered in his notebook to *Large Glass*, bonds for his system of playing roulette in Monte Carlo, *Three Standard Stoppages*, which he calls 'canned chance' – the list goes on. Moreover, as proved by Linda Henderson, Duchamp's thinking was inspired by Henri Poincaré's writings on the epistemology of science, which included important thoughts on the calculus of probabilities and its importance for the bases of scientific knowledge. In my analysis of Duchamp's art, I am less interested in analysing how this interest affected specific artworks than in his broader understanding of art as an intellectual game. I want to argue that his probabilistic approach to art – artwork as 'figuration of a possible' (1975: 73) and its exhibition as an intellectual endeavour to confound the viewer's expectations – were partially informed by his experience as a professional chess player, inasmuch it sensitised him to approaching art as a predictive operation. Moreover, his activities as an artist were not sustained for the sake of success in the game of art. In line with the surrealist aesthetico-political programme, he attached existential significance to art, defining it as exercise in freedom and exploration of novel modes of living, adjusted to the industrial environment of the twentieth century and its post-national culture. To elaborate more on the peculiarity of Duchamp's philosophy and artistic practice, I will conclude by turning to Robert Musil's *The Man Without Qualities*, comparing the French artist with Ulrich, the main character of the novel. I would like to argue that both characters – who blur the distinction between truth and fiction in opposite ways – embody a new kind of non-subjectivity centred on the notion of possibility. They are both rational eccentrics whose self-awareness is decisively formed by modern science and who sacrifice their integrity and identity in exchange for boundless possibility to do utterly improbable things.

In many ways, Duchamp's artistic path, despite similar interests and intuitions, leads in the opposite direction to that of the Futurists. If Marinetti and his cohorts praised acceleration and being intoxicated by the complexities of modern life, Duchamp sought spaces of temporary refuge from the numbing practicality of a middle-class existence. Clearly drawn to the idea of playing frivolous games, he abstained from the serious ones (the pursuit of professional success and social recognition). In 1911, one year before Boccioni completed his most sophisticated artistic and theoretical statement, *States of Mind*, Marcel Duchamp painted a picture on a very similar topic – *Nude (Study), Sad Young Man on a Train* (their correspondence was almost certainly accidental). Executed in a particularly cubist style, Duchamp's picture depicts a face fragmented into a row of horizontal blocks of shadow and light, sharing a very Rembrandtesque palette of colours and a certain heaviness of substance. Despite the fact that both painters shared a typically avant-garde distaste toward the rules of classical representation and chose a similar subject, and a very particular one, to say the least – the feeling of sadness prompted by train travel – they also clearly expressed very different attitudes and approaches. If Boccioni was drawn toward exploring the grey matter and the vagueness of forms ('gestalts') emerging in perception, then Duchamp dissected and analysed his subject with little regard for his emotions. His painting is sterile, geometrical and static. Octavio Paz rightly noted that 'Right from the start Duchamp set up a vertigo of delay in opposition to the vertigo of acceleration' (2011). The opaqueness of his painting – if not for the title it would be barely possible to recognise a human head, let alone a melancholy commuter – stems from the fact that he abstracts the scene into so many blocks of different shades of browns and yellows that it actually becomes a challenge to see past them and recognise or imagine a sight we know. If traces of a techno-scientific imagination were to be found in its style, then it would be non-Euclidean geometry and chromatographic deconstruction of movement, not Brownian movements or stochastic models of the unconscious, as with the Futurists. However, Duchamp's ways of thinking about and making art also developed in crucial, yet more subtle, directions with deep affinities to the probabilistic world-image. His interest with probability was divorced from the context of bodily sense perception. When asked by Pierre Cabanne what he thought of them, he answered:

'The Futurists, for me, are urban Impressionists who make impressions of the city rather than of the countryside' (1979: 35). As in many other cases, his argument against contemporary painting was that it was too 'retinal', that is, it engaged too much with the mundane activity of seeing the world. For Duchamp, art needed to be a cerebral process, involving philosophical deliberation, mocking scientific procedures, strategic planning, role-playing, recognising the social ramifications of art only to disrupt them, making more or less sophisticated jokes, and playing with words and images to construct multidimensional labyrinths of meanings, sometimes all at once. He saw himself as a typically French Cartesian spirit who doubts everything and renounces sensory knowledge (Ashton 1966: 244). If the Futurists worked on probabilistic 'sensibilities', then Duchamp can be accredited with investigating 'probabilistic rationality' through his works of art.

It is likely Duchamp's provocative intellectualism that won him a top spot on academia's shortlist of artists for writing about. Almost every conceivable aspect of his life and work has been thoroughly discussed: his alleged interests in alchemy, his investigations into human desire and eroticism, his interest in modern science, his career as a chess player, etc. Even his 'Apparently Marginal Activities' could not go unnoticed (Filipovic 2016). After over sixty years of critical attention and historical analyses, Duchamp seems to stand stripped bare by the art historians. It seems even his naked body has been taken apart and then reassembled over and over again to fit into different narratives about the tumultuous transitions of art discourse in the twentieth century. Despite being obviously 'detached' from the norms and habits of society, Duchamp figures in countless philosophical and art-theoretical texts as an artist whose work and life embodied cultural shifts on a larger scale. Depending on the point of view, he symbolises the emergence of 'anti-art', the transition from modernity to postmodernity, or the crisis of beauty as the dominant aesthetic value.[1] Frankly, I intend to proceed in a similar manner, and make use of Duchamp to make a point about the emergence of probabilistic subjectivity and the conditions under which it is possible. I will be arguing that Duchamp was shaped by the modernist literary culture of the late nineteenth century, but through his artistic and existential

---

[1] I also try to acknowledge throughout this study that Duchamp's practice embodied and elaborated its own theoretical positions, thus avoiding any facile application of external formulations that might otherwise divert our endeavour to bring into focus his art and the larger cultural and historical environment in which it operated.

experimentation arrived at a completely different aesthetic and epistemic paradigm, distinctly different from that of his contemporaries, even those belonging to the same artistic and intellectual circles.

My depiction of Duchamp as an artist whose 'chance operations' – a term he himself coined – were not intended to carry meanings or express definite goals (political, social, aesthetic, etc.) but play with art to achieve improbable results, rests on the recognition that his bias toward the unexpected was only possible within a peculiar understanding of one's own existence. It would probably be wrong (and unprovable) to assert that his life choices and artistic endeavours were calculated to prompt specific reactions or motivated by philosophical considerations, though his scattered writings and interviews demonstrate the self-awareness of a player who views art in terms of games and strategies. In my reading, I will be stressing the zaniness of his thinking, which, in my view, should always be regarded as an intellectual game as much about communicating ideas as it is about surprising others, or even himself. However, accentuating the playful aspect and intention should not overshadow its sincerity and profundity. In other words, I want to take his fondness for the non-serious seriously – as a positive existential project and a genuine quest to make a life more interesting and free. For example, when Duchamp was asked about his reasons for studying physics, a very serious task in itself, he replied that – contrary to common practice – 'it would be more of a game', making life 'more worth living' (Tomkins 2013: 85). His motivation behind engaging with the natural sciences was not to ground his method of practising art in some other discourse, as the Futurists did, but to liberate himself from the worn-out set of aesthetics, motifs and categories dominating the art discourse at the time. He cut and pasted scientific texts into poetic commentaries about his artworks, thus decontextualising both science and art in the process. Paradoxically, referring to physical laws, most notably in *Three Standard Stoppages*, 'a joke about the meter'[2] and other measure units, and *The Bride Stripped Bare by Her Bachelors, Even*, a convoluted visual essay on the nature of eroticism, he oscillated between serious (and pretentious) discussion and zany babble. This idea of using science against the narrowness of art also presupposed a strictly probabilistic understanding of freedom – not as the capacity to control one's life or realise one's goals, but as a wide array of possibilities to choose from, for doing the unexpected.

---

[2] From the Museum of Modern Art questionnaire about the *Three Standard Stoppages*, Artist's Files, undated but according to Naumann, written shortly after the acquisition of this work by the museum; see Naumann 1987.

By arguing for the essential non-seriousness of Duchamp's endeavours and, at the same time, taking them seriously, I also want to defamiliarise his image as an educated individualist who helped create the conditions for the institutionalised and highly academic discourse of contemporary art. Such a depiction of Duchamp, formulated exhaustively by Thierry de Duve in his impressive *Kant after Duchamp*, presents the artist as a revolutionary figure who is nonetheless firmly rooted in the core values of Western culture. While there is no denying that Duchamp's artistic path was in many ways shaped by his upbringing in a wealthy middle-class family, and then by the creative environment of Paris, it is also no coincidence that he distanced himself from every artistic or political movement and became truly recognised by the critics after his creative days were long past. I want to disagree with this point of view, though not for the sake of uncovering a deeper truth about Duchamp's inner motivations. That would be an absurd undertaking and of very limited benefit to the contemporary reader. Instead of trying to fit Duchamp into some historically significant tendency, I want to argue for his radical 'out-of-placeness': both in his own times and, even more, in the decades to come. (Can one even imagine a scenario in which Duchamp regularly goes for artistic residences and writes grant proposals for indefinitely unfinished works?) Duchamp's zany individualism did not fit particularly well on the Old Continent, nor in the hyper-individualistic culture on the other side of the Atlantic. Octavio Paz reported:

> His friend Roché has compared him with Diogenes, and the comparison is correct. Like the cynic philosopher and like all of the very limited number of men who have dared to be free, Duchamp is a clown. Freedom is not knowledge but what one has become after knowledge. It is a state of mind that not only admits contradiction but seeks it out for its nourishment and as a foundation. (2011)

Even if on the surface Duchamp presented himself as a radical individualist, almost a hermit living among the crowd in the biggest cities of the industrial age, he did not fit into any shared model of liberal or neoliberal subjectivity. His experimental existence did not really conform to the standard values of modern society, such as autonomy, freedom of choice, rationality, reliance on institutional order, or self-realisation (see Demos 2007: 6–8). By adopting fake identities (Rrose), suddenly jettisoning promising careers (for example, as a cubist painter), wandering back and forth from one continent to another (he travelled from Paris to New York to Buenos Aires, etc.), abandoning art for chess, devising dubious financial systems for winning at roulette, he was, in fact, playing with the idea of an autonomous and self-determining identity. His choices in art and life manifested a seemingly self-conflicting but actually quite strict and consistent logic which can be best explained in terms of a probabilistic self-awareness. Duchamp's lifelong quest to elude any kind of habit was an attempt at defying the seemingly natural tendency to seek self-confirmation and predictability.

Taming Chance (in Cans, Boxes, etc.)

Highly critical toward retinal art, which he saw as too focused on the sensory perceptions, Duchamp considered his paintings, art objects and performances to be platforms for wider aesthetic and epistemological explorations. Not only did he strive to introduce 'the precise and exact aspect of science' into art (Duchamp in Cabanne 1979: 39), though never for the same reasons as scientists, he also reflected on topics that were all the rage in the popular science discourse in the early twentieth century. He confessed he was never 'the scientific type', capable of using science constructively. His writings on physics, chemistry or mathematics were too hectic and inconsistent to receive any serious attention from the scientists or philosophers of his time. His method owed much to the Romantic tradition of irony: at once enthusiastic and detached, sensitive to the poetic aspect of language, even at its most precise and exact, and content with finding paradoxes which were never intended to be resolved. Among the many concepts he borrowed from modern science, two notions were central to his epistemological investigations – the fourth dimension and chance. Why these? Probably because they served as good points of departure for questioning and undermining fundamental assumptions about the true nature of reality: Euclidean geometry and determinism. My presentation will only focus on the notion of chance, discussing three works by Duchamp which explicitly relate to the probabilistic discourses of modern science. The first two, an unusual painting called *Three Standard Stoppages* (1913–14), consisting of three threads attached to a canvas, and *The Monte Carlo Bonds* (1924), a strange conceptual piece in the form of a wagering system and legal documents, I want to interpret as theoretical pieces demonstrating peculiar methods of applying chance procedures. The third, Duchamp's notes on his iconic *The Bride Stripped Bare by Her Bachelors, Even*, collected in a book titled *The Green Box*, will allow me to present his techniques within the larger frame of his philosophical explorations, revolving around the notion of possibility.

As noted by numerous scholars (Henderson 2005, Adcock 1984, Molderings 2010), Duchamp was introduced to the philosophy of science through the popular books of Henri Poincaré and all of the aforementioned works can be directly linked to specific passages in Poincaré's writings. Although at first the French polymath displayed a rather conservative view of the significance of the probability calculus he inherited from the classical thinkers like Pierre-Simon Laplace, he later came to acknowledge that chance plays a fundamental and revolutionary role in the emerging discourses of new physical sciences dealing with the complexity of fortuitous phenomena. For Poincaré, probability was not only 'the opposite of certainty', secondary knowledge useful only when more exact measuring equipment was lacking, but a necessary tool for predicting the behaviour of all systems sensitive to slight changes of initial conditions. In *Science and Hypothesis*, a book with which Duchamp was undoubtedly familiar (Adcock 1984), Poincaré gave three examples of such systems. The first one was weather, which, to this day, can only be predicted to a certain extent, with no indication that things might drastically change in the future. To illustrate this point, Poincaré points out that storms and cyclones occur in places where weather is particularly unstable and even the smallest events, 'a very slight tremor, or a breath of air' (2018: 398), might lead to disproportionate effects. The same rule applies even in the most elegant of the physical sciences, astronomy, which deals mostly with the very predictable behaviour of enormous objects like stars, planets or galaxies. He argues that the positions and paths of planets, now perfectly predictable, were initially beset by minor perturbations, as evidenced by the random distribution of their orbits. The final example, and the one which captured Duchamp's imagination the most, concerned the unpredictable movement of a different sphere – a ball tossed onto the roulette wheel. Here, the gambler's fate depends on events that are beyond human perception and control – the initial spin of the ball and the wheel. In all of these cases, chance is understood both subjectively, as insufficient knowledge, and objectively, as a multiplicity of small causes producing disproportionately large effects. This leads Poincaré to believe that '[f]or the fortuitous phenomena themselves, it is clear that the information given us by the calculus of probabilities will not cease to be true upon the day when these phenomena shall be better known' (2014: 397). By their nature, scientific theories address only certain aspects of material events and thus cannot take into account all the factors that might shape their development. To make perfect predictions of the weather one would need to take into account movement patterns of butterflies and birds, the frequency and intensity of the respiration of all living creatures, etc. And these factors also depend on a multitude of causes (like the weather itself).

Poincaré's perspective on chance and probability were consistent with his conventionalist understanding of science as 'a system of relations' (350). In his view, mathematical systems shared across all the sciences, such as geometry, should not be regarded as *a priori* truths, as it is possible to invent different (non-Euclidean) geometries and use them felicitously or infelicitously in various contexts. He wrote:

> The axioms of geometry therefore are neither synthetic a priori judgments nor experimental facts.
>
> They are conventions; our choice among all possible conventions is guided by experimental facts; but it remains free and is limited only by the necessity of avoiding all contradiction. Thus it is that the postulates can remain rigorously true even though the experimental laws which have determined their adoption are only approximative.
>   In other words, the axioms of geometry (. . .) are merely disguised definitions. (65)

As no theory is made of true representations of reality, but of inexorably imperfect artificial models of physical processes, applicable only under certain circumstances, chance, occupying the gaps (or 'stoppages') between different scales and systems of relations, cannot be abolished. Duchamp's *Three Standard Stoppages* comments on both of Poincaré's ideas, his understanding of chance and of the conventionality of scientific theories, which inspired not only its title, but also its form and the creative method (Molderings 2010: 42–4). It was made of three separate pieces of long and narrow glass covered with strips of canvas. Onto this background Duchamp attached three pieces of thread, each approximately one metre long, and then placed all the pieces in a wooden croquet box. The shape and placement of the threads were determined experimentally: Duchamp dropped them onto the canvases from a height of one metre, allowing the imperceptible movements of air to decide how they would hit the ground. The final effect was preserved in varnish.

Duchamp termed the effect of his modest experiment 'canned chance', meaning it was an attempt at capturing a fortuitous event as a work of art. To accentuate and simultaneously ridicule the ritualistic aspect of his gesture – the transformation of the threads into art objects – he encased them in three frames, one stacked over another. First, he placed them on canvas, traditionally a painter's material, then he fixed the canvas on a glass panel, and finally enclosed the whole set in a big wooden box.[3] In this sense, he represented and

---

[3] The pieces are displayed separately in the museum.

objectified chance – an event of which he could not have knowledge or control – by literally putting its material result in a box and signing it with his name (just as he appropriated a urinal and submitted it for an art exhibition as the *Fountain*). However, the aesthetic transformation of mundane objects or events was not his sole intention and, as usual, he hinted at a few other possible interpretations of his peculiar gesture. Above all, in contrast to other artists who, around the same time, also experimented with randomness, such as Hans Arp, he did not use chance for purely aesthetic purposes. The very gesture of enclosing the threads in multiple boxes might be seen as a reference to laboratory practices. In many analyses of Duchamp's art, *Three Standard Stoppages* is considered a critique of standardisation, and the three threads represent unique and individual, though ironically 'standard', manifestations of a metre whose true standard was also preserved in a box and presented as a scientific artefact in Sèvres. Moreover, its title refers directly to Poincaré's critique of these idealistic scientists, called the 'Thread School' in *Science and Hypothesis*, who take mathematical abstractions too seriously. In this sense, the painting is a joke, an amusing exercise in physics, conceived and executed in the spirit of Alfred Jarry's pataphysics, and directed against the fetishisation of standard measures among serious people who identify as rationalists.

Finally, *Three Standard Stoppages* can be interpreted in yet another way: as an early example of using random processes in art (the physical act of dropping threads on the ground) to undermine the notion of absolute authorship. By granting agency to chance, Duchamp makes it apparent that the final effect on the canvas bears only a probabilistic relation to the artist's intentions. In other words, as the artist, he can only take partial responsibility for his creation. This interpretation is consistent with Duchamp's later statements, in which he declared that his art is not made to please the general public, just himself, and this is only possible when the effect succeeds in surprising and amusing him. Relinquishing control over the creative process in its totality allows him to enjoy his own art as a curious spectator who is never sure what is lurking around the next corner. It is no coincidence that, in conversations with Cabanne, the word 'amusing' appears almost every time Duchamp speaks about the ideas behind his artworks, as if 'amusement' was a crucial component in his artistic practice. Canning chance is just one of many activities he found amusing for their straightforwardness and unexpected results (47). It is also no accident that the word 'sublime', a key concept in the vocabulary of Western aesthetics, does not appear in those conversations even once, even though Salvador Dalí manages to squeeze it into his short Preface twice. When contrasted with Duchamp's witty but reserved tone, Dalí's ecstatic eulogy exposes Duchamp's peculiarity as an avant-garde artist who, in contrast to the Spanish megalomaniac, understands art not as an expression of individuality but as an amusing game of chance. This playful attitude corresponds with a deeply probabilistic world-image. As Duchamp himself clarified, during work on *Three Standard Stoppages* his intention was to 'forget the hand' – his craft and personality – because 'fundamentally, even your hand is chance' (1971: 46).

Making Art at the Roulette Table

In Duchamp's own words, *Three Standard Stoppages* was a painting *of* chance. By comparison, another work of art that dealt explicitly with chance and probability, *The Monte Carlo Bonds*, was described by him as a drawing *on* chance. If the concept of the former rested on the intention to capture chance on canvas, thus translating a random event into a static image, the latter can be considered in terms of a complex performance, in which randomness quantified as probability was the very substance of the art. Given its complexity and elusiveness as an art work, *Monte Carlo Bonds* figures among the least discussed of Duchamp's works, even though it demonstrates an understanding of art as a financial instrument that was unique for its time (with the notable exceptions of Damisch 1979, Joselit 1992 and Judovitz 1995). Although it did not resemble any existing traditional, or even avant-garde art genre, the project was promoted by the press as an art event and a financial opportunity for art collectors. Importantly, this presentation was intended by Duchamp, who asked his friend Jane Heap, the editor of *The Little Review*, an American literary magazine, to publish a short announcement about his endeavour in the fall/winter issue from 1924. She wrote:

> Marcel Duchamp has formed a stock company of which he is the Administrator, etc. Shares are being sold at 500 francs. The money will be used to play a system in Monte Carlo. Stockholders to receive 20 per cent interest, etc. Some of the shares have arrived in this country and are very amusing in make-up. They carry a roulette wheel with a devil-like photograph of Marcel pasted upon it, they are signed twice by hand, – Prose [sic] Sélavy (a name by which Marcel is as well-known as by his regular name) appears as president of the company. If anyone is in the business of buying art curiosities as an investment, here is a chance to invest in a perfect masterpiece. Marcel's signature alone is worth much more than the 500 francs asked for the share. Marcel has given up on painting entirely and

has devoted most of his time to chess in the last few years. He will go to Monte Carlo early in January to begin the operation of his new company. (In Duchamp 1975: 185)

While Heap's description focuses on the quality of the bond as an aesthetic object, the visual aspect was secondary to the project's performative dimension. This was initiated by forming a joint-stock company board in which Duchamp acted as the administrator and Rrose Sélavy, his feminine alter ego, was nominated president. The company claimed to have found a roulette system which would bring investors annual income. It relied on a painstakingly boring process of throwing dice to place bets and adjusting the amount of money wagered to slowly accumulate winnings. The specifics of the system were never revealed, but it was a type of martingale – a betting strategy popularised in France in the seventeenth century. Its simplest example was designed for the game of heads or tails, and revolved around doubling bets after every loss so that the next win would compensate for previous losses and make the profit equal to

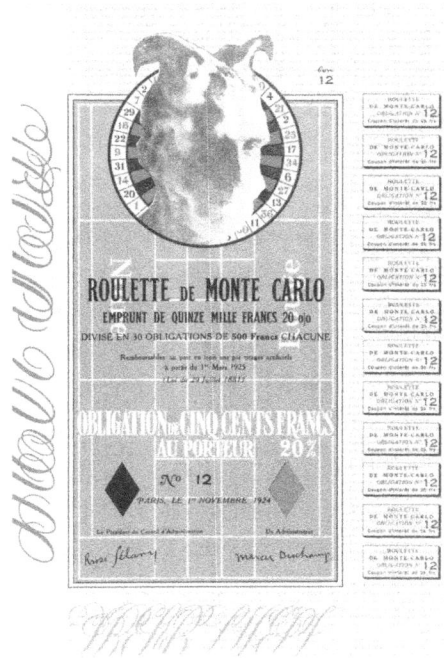

Figure 4.1 Marcel Duchamp, *Monte Carlo Bond* (No. 12). Copyright © 2023 Artists Rights Society (ARS), New York / ADAGP, Paris / Estate of Marcel Duchamp

the original stake. The rationale behind this strategy was flawed, because it required the gambler to possess infinite resources to continue placing bets. As the bets grew exponentially, the gambler eventually had to go bankrupt. Duchamp was clearly unaware of this fact, and in his defence, it should be noted that the possibility of finding the perfect martingale was mathematically disproved in 1934, ten years after *Monte Carlo Bonds* were issued. Moreover, in all probability, Duchamp's system was more complex than the one described above. He had been working on it for a long time in his Paris apartment, establishing an amateur laboratory of probabilistic praxis, before giving it a spin at a real roulette table in Monte Carlo where gravity, 'the ministry of coincidences' prevails (Duchamp 1975: 33). What can be held against Duchamp is that, in a letter to the art collector Jacques Doucet, he boldly proclaimed, probably under the influence of gambling euphoria, that: '[T]his time I believe I have eliminated the word chance' (1975: 188).

There is thus some merit to Dalia Judovitz's interpretation that *Monte Carlo Bonds* 'may be considered as yet another instance of Duchamp's efforts to "can chance"' (1995), or Hubert Damisch's view that the martingale was 'intended to counteract chance, to trick it, catching it up in its own laws, to treat it finally as an opponent' (1979: 19). Yet I would argue that there is a significant difference between capturing chance in the form of an object and taming it with a probabilistic system intended to operate infinitely and maximise the value of the art collector's investment. The uniqueness of this endeavour lies in its elusive product. Duchamp's experiments with randomness did not materialise as an artefact, other than the bond itself, but offered a potentially infinite source of income. As a failed experiment in 'counteracting chance', *Monte Carlo Bonds* are a mere curiosity in art history; however, when read in context of his interest in the logic of the art market and his much later reflections on the random nature of success in art, they come across as fascinating attempts at equating art with financial speculation (which is ultimately a game of chance). Buying the bonds, like any other art object in the capitalist context, was an investment shrouded in uncertainty. In a letter of 1952 to his sister, Suzanne, and her husband, Jean Croti, he wrote: 'Artists of all times are like gamblers of Monte Carlo, and this blind lottery allows some to succeed and ruins others. In my opinion, neither the winners or losers are worth worrying about. (. . .) Everything happens through pure luck. (. . .) And even posterity is a real bitch who cheats some, reinstates others (El Greco) and reserves the right to change her mind every now and then' (1982: 16–17). If '[m]illions of artists create' and 'only a few thousand are discussed or accepted by the spectator' (1975: 138), as he states in *The Creative Act*, then neither prioritising self-expression nor communication with the audience is a better strategy than the other. Duchamp looks at art history much as Charles Darwin looked at evolution – as an enormously complex game in which success depends on many variables and is statistically improbable. The artist is never alone in their explorations, because – consciously or not – they always draw on previous artworks and as soon as they present their own work to the public, it enters through the galleries and museums, the vast databases of cultural heritage, and is valorised in a context. It seems that after almost thirty years, he gave up on his ambition to conquer chance and settled on a probabilistic standpoint by which all art was a form of gambling.

Talking about the project almost forty years later, Duchamp maintained that winning at the roulette table was never his concern, as he was always more interested in the 'intellectual side' of his part-gambling, part-mathematical (and all-artistic) experiment. In a television interview with James Sweeney for NBC, he stated:

> As you know, I like to look at the intellectual side of things, but I don't like the word 'intellect.' For me intellect is too dry a word, too inexpressive. I like the word 'belief.' In general when people say 'I know,' they don't know, they believe. Well, for my part, I believe that art is the only form of activity in which man, as man, shows himself to be a true individual who is capable of going beyond the animal state. Art is an outlet toward regions which are not ruled by time and space. To live is to believe, that's my belief.

For Duchamp, the intellect did not clinically operate on concepts to illuminate one's path to immutable truths. It was always, to some extent, a form of gambling in itself: making surprising or even absurd conjectures and testing them with the greatest patience. His epistemic scepticism, or rather, affirmation of uncertainty, was not necessarily a manifestation of the intellectual trends of his day. His fondness for Poincaré's epistemology, and later, Ludwig Wittgenstein's philosophy of language, fits well within and expands upon his self-proclaimed intellectual 'Frenchness' which he perceived, somewhat peculiarly, as an attitude toward reality that wavers between analytic exactness and doubtful hesitance. In an interview with Dore Ashton, Duchamp confessed that he happened 'to have been born a Cartesian. The French education is based on a sequence of strict logic' (Ashton 1966: 244).[4]

---

[4] During the great theological debates of the thirteenth century, Stephen Tempier, the Bishop of Paris, condemned philosophical positions rooted in Aristotelianism on the ground that God's infinite power cannot be grasped through the principles of logic. A similar position was shared by Blaise Pascal, who famously argued against the possibility of grounding faith in reason, accentuating the epistemic urgency of doubt.

And although he claimed to have never read Descartes's treatises, he felt deeply influenced by Cartesianism and its methodical 'acceptance of all doubts' (245).

A Probabilistic Eccentric

However, there were significant differences between Duchamp's modest epistemology and that of the great French philosopher. As famously expressed in his 'Cogito, ergo sum', doubt grounded Descartes's notion of a fixed subject which must exist beyond any doubt, as it confirms itself by being able to doubt in the first place. By comparison, Duchamp's natural inclination toward doubt and logical exactitude went hand in hand with his aesthetic and epistemic rejection of strong subjectivity (which every avant-garde movement vilified for being a backbone of bourgeois psychology). As a result, his personal take on the 'strict logic' of Cartesianism was very much marked by his sense of humour and strong inclination toward clownish behaviour. Many of his activities, like manufacturing his own wanted poster, would be best described as Diogenean in spirit – an attitude that hardly fits within the tradition of analytic rigour and discipline to which Duchamp, not without cause, claimed to adhere.

It is, however, important to note that, in the social circles of Bohemian eccentrics in Paris, Duchamp's bias toward self-contradiction did not particularly stand out. For example, in his early days in Paris he became friends with Francis Picabia and Guillaume Apollinaire, a duo of tricksters at least as defiant as him. Gabrielle Buffet-Picabia recalls that her husband and Duchamp:

> emulated one another in their extraordinary adherence to paradoxical, destructive principles, in their blasphemies and inhumanities which were directed not only against the old myths of art, but against all the foundations of life in general. . . . Better than by any rational method, they thus pursued the disintegration of the concept of art, substituting a personal dynamism . . . for the codified values of formal Beauty. (In Tomkins 1972: 31–2)

It was in the very spirit of French 'dandyism' to provoke middle-class tastes and challenge shared values and lifestyles. Jules Barbey d'Aurevilly, a late Romantic novelist who greatly influenced *fin de siècle* Symbolists and Decadents, wrote that 'one of the consequences of Dandyism, one of its principal characteristics – or rather, its most general characteristic – is that of always producing the unexpected, that which the mind accustomed to the yoke of rules cannot logically expect' (in Gill 2009: 79). In other words, a penchant for the unexpected constituted the very essence of bohemian eccentricity. However, if one is to believe Peter Stallybrass and Allon White's theory of bourgeois identity, the eccentrics played a very important role in building the identity of the class they were revolting against. In their view, the nineteenth-century middle class defined itself by the rejection of the irrational and the eccentric, restricting the space of acceptable behaviour, opinions or public appearances. Conversely, by raising (and codifying) the standards of propriety and pushing various forms of otherness outside the social body, the bourgeoisie could also constitute its own identity. As Stallybrass and White argue: 'the radical democratic project was nothing more nor less than this process, in which a "natural", "middling", "democratic", "rational" subject was laboriously constructed by a rejection of all specific and particular domains' (1986: 199). The Bohemians – for various reasons and in many different ways – revolted against such a restrictive standard of subjectivity.

One such tactic against the middle-class limits on public expression of personality was the self-conscious and intentional fashioning of one's identity, for which Joséphin Péladan coined the term Kaloprosopia (from Greek *kalós*, 'beautiful', and *prósōpon,* 'face, person') in 1894. Péladan was describing a practice already common among Bohemian artists, that is, considering not only one's public image, but even personhood as one of many art forms which can be practised on an everyday basis through theatre, literature, fine arts or social behaviour, seen as small performances (Deak 1991). Among the most famous 'kaloprosophs' were Oscar Wilde, who famously admitted to caring more about fashioning his life into an artwork than his literary output, and Stéphane Mallarmé, whose public persona was as influential in French literary circles as his poems. Duchamp, an avid admirer of Mallarmé, considered both his public image and his body and mind to be media for avant-garde art throughout his artistic career. The aforementioned performances staring Rrose Sélavy are the best examples of this aspect of his art, though many other minor gestures demonstrated his eagerness to playfully experiment on himself in the name of art. One particularly amusing experiment took place aboard a ship on which Duchamp was returning from Argentina to France, where he cut a 'star-shaped tonsure' in his hair in homage to the 'comet with its tail at the front' and *Tête étoilée* (*Starry Head*), a poem by Apollinaire. Significantly, these minor eccentricities were never intended to simply provoke the general public, like similar gestures by his Dadaist and Surrealist friends, who more often than not sought to offend and enrage their audiences. In Duchamp's case, the minuteness and intellectualism of his gestures shows, rather, that his unique method of 'kaloprosopia' should be understood differently: not in terms of transgression (of cultural norms) but as an affirmation of the possibilities afforded by bourgeois life and its institutions. Contrary to the late nineteenth-century *décadents* or his fellow avant-gardists, Duchamp positions himself not as an antagonist of the bourgeois but as one of them, sabotaging the system of norms and limits society imposed on itself from within.

From this perspective, it becomes clear that Duchamp's quasi-scientific musings on chance should not be seen merely in terms of mocking the academic language. The preface to his collected writings accompanying *The Bride Stripped Bare* abounds with notions from a physics textbook:

> Given   1. the waterfall
>   2. the illuminating gas,
>
> one will determine
> we shall determine the conditions
> for the instantaneous State of Rest (or allegorical appearance)
> of a succession [of a group] of various facts
> seeming to necessitate each other
> under certain laws, in order to isolate the sign
> of [the] accordance between, on the one hand,
> this State of Rest (capable of [all the] innumerable (?) eccentricities)
> and, on the other, a choice of Possibilities
> authorized by these laws and also
> determining them. (1975: 27–8)

Obviously, Duchamp's erudite but convoluted introduction does not allow us to determine a clear-cut meaning, primarily because even the subject of this paragraph remains vague and is thus open to many equiprobable interpretations. Only one thing is evident: its lexicon, unusual for a text concerned with a work of art, is abundant with words which would normally appear in a treatise on fundamental physical laws. Its crucial notion is 'possibility', which reappears several lines later in the 'algebraic comparison', translating the quoted fragment into a concise formula with two parameters: 'a', above the bar ('being the exposition'); and 'b' ('being the possibilities'). The nature of this equation, whose result remains unspecified, is statistical or probabilistic, as it implies considering art as in terms of coding, or as operating on possibilities. The ratio is not equal to anything specific, as it applies both to the act of creation – the derivation of an actual artwork from the vast pool of possibilities – and reception – the process of its decoding by a viewer whose perception is limited to the context of his expectations.

The lack of an 'equals sign' indicating the result also seems to suggest that art should be interpreted as an abstract process that takes place in the virtual space of possibility. This reading is consistent with Duchamp's appreciation of cerebral (non-retinal) art, which should be freed from expressive or mimetic obligations. Herbert Molderings rightly argues that the notion of possibility is key to Duchamp's aesthetic project:

> 'Likeness' and 'truth' were not its key aspects, as in all the various brands of realism, were nor beauty, taste, and harmony, as in the aesthetics of formalism, but rather 'the possible,' in the sense of what is merely conceivable, the idea that all things can be perceived and conceived differently. (2010: 130)

Duchamp rejected likeness and truth as unnecessary and purely conventional determinations, which he called 'callistics', the 'science of the beautiful', which needlessly narrows the field of possibility for artistic creativity. On the one hand, his explanation for abandoning the old aesthetic categories fitted squarely into the cultural climate of nineteenth-century Romanticism, when the idea of art for art's sake (*l'art pour l'art*) was first expressed, and was widely embraced by Bohemian social outsiders. On the other, the Bohemians, who advanced and cherished the idea of art free of utilitarian constrictions, convinced that individuality was of the highest value, also became crucial in promoting the concept of personal freedom as the ultimate truth of human existence – a philosophical and political standpoint Duchamp did not share in the slightest. While it is true that Duchamp valued absolute individuality over any shared notion of subjectivity, his approach to the subject was strictly apolitical and thus non-liberal: as a locus of exceptionality and quaintness rather than of a strong identity. For that reason, he renounced realism and always expressed his contempt toward 'retinal painting', an ordinary portrayal of how one sees the world, in discussing his approach to art. In Duchamp's eyes, there was nothing interesting in witnessing the truth of someone else's perception. Unlike his peers, he saw Impressionism as a sign of deterioration in painting, too obsessed with the mundane reality of the individual existence.

Possibility Erases Identity

However, Duchamp's discontent with art focused solely on individual sensory perception did not lead him to embrace outmoded Symbolism or pure abstraction. Instead of invoking metaphysical constructs, he centred his aesthetic explorations around the notion of 'the possible', understood quite instrumentally as 'caustic', an artistic material that acts upon existing forms and rules in art, in Duchamp's own words, 'burning up all aesthetics or callistics' (1975: 73). This highly metaphorical expression translates quite well into useful instructions on how to practise art. Duchamp's realm of possibility is an ever-elusive attractor in the artist's mind, 'the other condition', engaged to free oneself from conscious or unconscious determinations and the urge for self-confirmation. In simpler terms, it is an aesthetic method which allows one to imagine and perceive things in an unusual context. Unlike the realists, who aimed at modelling perceptions, characters or events that the audience may find probable, the Duchampian possibilist focuses on enriching the fabric of perception and cognition with alternate and mostly improbable ideas, meanings or behaviours. However, if Duchamp expresses the general idea of the 'possibilistic' art to avoid self-contradiction, he never discloses what this 'possible' is, precisely: the possible is NOT 'the opposite of impossible, nor as related to probable, nor as subordinated to likely'. However, from this negative definition we can still induce that Duchampian possibility is quite synonymous with unlikeliness (or improbability). Hence, the function of art seems almost synonymous with entropy, understood as an increase of epistemic uncertainty. Looking for improbable uses for objects, finding unlikely connections between things and ideas, mixing up different discourses and cultures, all to enchant the world (so that ordinary things are no longer what they seemed). All this 'weird' behaviour also has an entropic effect on art and theoretical discourse. The possible manifestations of art grow exponentially, not unlike the theoretical discourses, which were put there in the first place to systematise and thus reduce the uncertainty around the multitude of existing art practices (a futile project, given the open-endedness of Duchamp's gestures poking holes in cultural systems of producing, classifying and evaluating art). Like viruses which mutate and replicate inside the system to change its rules and logic from within, their intended effect is the expansion of possibility: of what is conceivable as art, but also of the possible meanings and associations evoked by everyday objects. Almost every *ready-made* can be explained in such terms: the very first one reinvents the stool as a possible base for a wheel (it was meant as a practical object – the wheel could be spun and contemplated like 'the flames dancing in a fireplace' (in Schwartz 1969: 442)), and at the same time introduces bicycle wheels and wooden stools into the set of possible art media. This is art made by addition – of an abstract dimension onto the uncompromising practicality of the three-dimensional world of objects and causal rationality – instead of subtraction – of multidimensional experiences on two-dimensional surfaces of a canvas like in the case of 'retinal' arts.

Duchamp's explanation of the creation and reception of art as a cerebral process relying on the vastness of possibility space also implies excluding the authority of the creator (and their intentions) from the equation, because their virtual presence narrows the interpretive possibilities. Following the early observations collected in *The Green Box*, in *The Creative Act* Duchamp argues that the artist is 'a mediumistic being' (1975: 138) who 'plays no role at all in the judgment of his own work' (139). In a sense, exactly a decade before Roland Barthes published his famous essay 'The Death of the Author' (1982), Duchamp had already internalised its most important assumption, that the artist's identity is a supposition which sustains the tyranny of the one and only true interpretation. However, for Duchamp, it is not only the recipients – especially students of literary theory and art history – who should spare themselves the trouble of uncovering the true intentions behind a work, the artist is invariably removed from their own creation. I want to argue that Duchamp's reliance on chance allows us to consider this equation in probabilistic terms, as the intentional inclusion of randomness in the creative act, as in *Three Standard Stoppages*, revealing and openly exhibiting that the relation between the artist and their work is irreducibly uncertain.

Duchamp borrows the idea of an artist as medium from T. S. Eliot, whom he quotes in *The Creative Act*, but goes beyond his predecessor's scope of interest. He refers to Eliot's seminal essay 'Tradition and the Individual Talent' (1919), which lays down his 'Impersonal Theory' of poetry. Therein, he compares the author to a chemical agent which serves as catalyst in a reaction, who merely mediates in the process of exposing vast databases of cultural code to contemporary reality. The poet does not create from scratch, but engages in a centuries-long process of communication, leaving traces in the form of poems, melodies, images, etc. Like Duchamp, Eliot highlights the impersonal nature of this process:

> The analogy was that of the catalyst. When the two gases previously mentioned are mixed in the presence of filament of platinum, they form sulphurous acid. This combination takes place only if the platinum is present; nevertheless the newly formed acid contains no trace of platinum, and the platinum itself is apparently unaffected: has remained inert, neutral, and unchanged. (2014: 109)

In Eliot's theory, there is no room for such sacred notions of the Romantic tradition as self-expression, originality or (divine) inspiration. Creation is a 'continual surrender of himself' (108) to the immense archives of tradition which determine the possibilities for expression.

To this idea, which was already radical in its anti-humanism, Duchamp adds another layer of meaning. Even if Eliot maintains that writing poetry involves sacrificing one's personality, he still considers art in terms of authors and spectators. By contrast, Duchamp consistently avoids this dualism and proposes viewing art in relativistic terms: as a dispersed and distributed process of communication. When he argues that 'the creative act is not performed by the artist alone', he means this quite literally: the work of art does not exist objectively, but rather re-emerges continuously as it is viewed, interpreted and valued by society. The spectator is just as necessary in the catalysis process – the transfiguration of piece as the author, because he 'brings the work in contact with the external world by deciphering and interpreting its inner qualification and thus adds his contribution to the creative act' (1975: 140). In Duchamp's view, our perception of an artwork is determined by knowledge and expectations, becomes entangled with it, and cannot be extracted from the context. This leads Duchamp to assert two things. Firstly, it is impossible to forget that one is looking at a painting by Leonardo, for example, once one has learned this fact. The caption under the picture becomes an integral part of how it is perceived. Secondly, the aesthetic experience in front of a painting is never individual, because the spectator's gaze is decisively shaped by others' interests and opinions. 'The poor Mona Lisa is gone because no matter how wonderful her smile may be, it's been looked at so much that the smile has disappeared. I believe that when a million people look at a painting, they change the thing by looking alone' (Tomkins 2013: 60–1). Following this argument, it can be said that millions of tourists flocking to see Leonardo's masterpiece constitute the painting's (statistical) value as a remarkable cultural phenomenon, adored or detested for the very same reason: its popularity. In 1911, Duchamp witnessed a great scandal surrounding the theft of *Mona Lisa* from the Louvre and the ensuing rise of the painting's popularity. Perhaps no other event in art history evidenced so strongly that art's value is socially constructed, and that the key to success on the art market lies in publicity. If this is true, the artwork – as a material object – is secondary to its position in social networks and weight as a piece of information. As a result, it does not suffice to make art. One also must play the game of art, like chess, 'constructing some mechanism of some kind by which you win or lose' (. . .) – like, for example, submitting a urinal for an exhibition (for a deep view into intricacies of this 'move' see Kilroy 2018: 47–73). Recognising that art is a social game among many others allowed him to see aesthetics norms as rules. These, in turn, could be used or even changed, exploiting the fact that, as de Duve rightly observes, 'consensus – in art as in other domains of social life – is always somewhat blurry and unreal; that it is never anything but a statistical distribution of opinions, bunching up around its mean but significant above all in its standard deviation' (1996: 15).

Unlike Andy Warhol, for instance, Duchamp did not attain mastery in the game, he never played to win. Even if, for the construction of aesthetic 'mechanisms', it is necessary to operate within rules that determine winning or losing, Duchamp contended that 'the competitive side of it has no importance. The thing itself is very very plastic. That is probably what attracted me in the game' (Nelson 1958: 89–99). Although he understood art as a social and financial game, he was interested in playing differently from others, in investigating the game and probing its boundaries. Nothing better exemplifies his attitude in this regard than his famous advice, given supposedly to John Cage: 'Don't just play your side of the game, play both sides' (Kostelanetz 2003: 12). Unlike a businessman such as Warhol, whose cold and cynical awareness of the art-game materialised in economic success, for some reason Duchamp abstained from seeing himself as a player. On the contrary, he stubbornly maintained his position as an artist, uninterested in economic matters, exercising his (relatively modest) social privilege to pursue freedom from the mundane facets of practical life (like the necessity of winning certain games to ensure one's survival). Clearly, Duchamp valued the experience of possibility – being free from identities, social roles or a set place in the hierarchy – over measurable forms of accomplishment. In this regard, he resembles a fictional character, Ulrich from Robert Musil's *Man without Qualities*, who was conceived at approximately the same time as Duchamp was creating his most important works, in the early 1920s. The two characters share a very interesting combination of beliefs and values, such as a fondness for exploring the possibilities afforded by the intellect, statistically motivated anti-humanism, and, most importantly, a desire to be purged of qualities. As noted by Jerrold Seigel, who compared Duchamp's and Musil's works in relation to the subject of identity, they were 'two well-known advocates of a fluid and unsettled manner of individual existence' (2009: 24), who witnessed and described the historical emergence of the modern concept of selfhood. Taking my cue from Seigel, I want to focus on their common preoccupation: experimenting with a self-aware mode of existence, attuned to randomness and inclined toward the most improbable events.

A 'Sense of Possibility'

In contrast to Duchamp, Musil led a relatively unadventurous and staid life for a person of his social background. Born in 1880 in Klagenfurt, Austria, he was the son of Alfred, an engineer and the chair of Mechanical Engineering at the German Technical University in Brünn (Brno), who 'was a rather shy, reserved man and was dominated by his wife, Hermine, a woman of nervous unpredictability' (15). After a short and disheartening attempt at becoming an officer, he returned home and enrolled as a student of mechanical engineering in his father's department. This was where he became acquainted with modern literature, an encounter which probably determined his decision to pursue further education in philosophy and psychology. For a couple of years he worked as a civil servant in various German-speaking cities (mostly Berlin and Vienna), finally devoting himself to work as a novelist, essayist, editor and theatre critic. *The Man Without Qualities* tells a similarly unremarkable story of a man who drops out of university, but instead of writing, like Musil himself, decides to devote his life to a grand aesthetic experiment. Thus, even though many clues allow us to establish biographical connections between Musil and the main character of his novel, Ulrich, *The Man Without Qualities* is more of a virtual playground for experimenting with improbable yet conceivable modes of existence than a biographical parable.

The narrative centre of the book is dislocated between two voices which provide the reader with philosophical, social and political commentary: Ulrich, and an elusive narrator whose knowledge about the world varies, depending on the occasion. Moreover, the fictional characters mostly engage in lengthy discussions and share observations which, in other novels, would be solely reserved for the narrator. Sometimes entire chapters consist of an inner monologue occasionally interrupted by the narrator's ironic commentaries; often chapters are linked through philosophical problems and not by the development of the story, which, incidentally, barely moves forward and is never resolved, as Musil died before completing the second part, leaving the end indefinitely open.[5] Notably, the book's

---

[5] However, it would be wrong to assume that the story evolves only to confirm some conceptual or philosophical argument. It is more complicated than that. The world of intellectual commentaries and speculations is often affected by even the tiniest events in the fictional world, suggesting that every thought is tied with its direct environment, which does not necessarily dictate its content and form but undoubtedly shapes it in some way.

incompleteness seems consistent with its philosophical preoccupation with indeterminacy (of physical events, personal fate or social reality) and its irreducibility to narrative explanations, even though Musil – unlike Duchamp who left *The Bride* . . . forever unfinished on purpose – never expressed an intent to leave his work open-ended. If in Duchamp's case the unfinished form of the work hinted at his inability to formulate a grand theory of eroticism, in *Man Without Qualities* the lack of closure expresses the main character's similar state of helplessness in his futile attempts to find a radically different way of life. In the novel, this is referred to as the Other Condition, beyond the rule of social norms and ordinary perceptions of reality where 'the boundaries between mind and world blur, the subject and the object fade into each other' (Di Bona and Ercolino 2019). As Reza Negrestani rightly observes, this idealised state 'is not flatly a rejection of reality and a withdrawal into the mystical, but rather, an expansion of other kinds of realism that include the non-expressible and non-rational conditions of experience' (Negrestani 2018). As in Futurist paintings (Boccioni) or Surrealist poetry (Breton), Musil's fiction is decisively shaped by post-Newtonian science and post-positivist epistemology, so that his take on 'the modern ineffable is experienced as induced from statistical probability (. . .) or objective conditions of multiplicity in the world – that include objects, systems and numbers – resulting in the reciprocal mutability between qualities and quantities' (Negrestani 2018).

Ulrich's quest is not psychologically motivated. As the title suggests, contrary to typical heroes of realistic fiction, Ulrich lacks character – or purposefully chooses to live as if he had none – which could explain the motivation for his actions. After giving up on socially accepted ambitions, such as achieving high social status (like Musil he drops out of the military) or intellectual success (as a mathematician), Ulrich decides to consciously create himself as a 'man without qualities' who abstains from a goal-oriented existence. The book focuses on his life in Vienna, to which he returns after abandoning a university career, a capricious decision prompted by an article in a newspaper in which a racehorse was declared a 'genius', signifying for Ulrich the ultimate trivialisation of traditional forms of social status. If a racehorse can be called a genius, this means that 'genius' is no longer absolutely exceptional and improbable. When his father presents him with a small palace in Vienna, he is afforded the time to indulge in the typical activities of a Viennese gentleman without the pressures of having to make a living. He starts by procuring himself a mistress, and engages as honorary secretary in a bizarre political project called 'The Collateral Campaign', a large-scale preparation for celebrating the seventy-year jubilee of the old emperor, Franz Joseph. However, Ulrich plays no significant role in the undertaking, which in itself remains undefined, as he takes part in endless (and pointless) discussions between members of the old elite and reflects upon events and situations unfolding before his eyes. He only gets involved in meaningless actions. While doing nothing useful or meaningful from a social perspective, he ponders and tries to enact an elusive mode of existence that does not define his identity or motivations for actions. Musil makes Ulrich's existential project more probable and realistic by setting his life in an environment where he can afford an idle existence. However, he does not stop at explaining it in materialist terms, as mere social entitlement. What steers Ulrich toward seeking an identity without qualities is less economic than epistemic, that is, belonging to an early twentieth-century state of knowledge. For the main character, a former scientist, even the most abstract and technical knowledge actively affects his self-image and subjectivity. Being motivated by a vague need to become exceptional – though this is never explained in psychological or psychoanalytic terms – he rejects all the ordinary paths to this goal which presume professional success and social recognition. A prominent place in the social hierarchy, economic prosperity and academic achievements are as worthless to him as belonging to the old humanist culture. The narrator elaborates on Ulrich's (im)personal problem in broader context of epistemic shifts occurring in Western culture:

> Probably the dissolution of the anthropocentric point of view, which for such a long time considered man to be at the center of the universe but which has been fading away for centuries, has finally arrived at the 'I' itself, for the belief that the most important thing about experience is the experiencing, or of action the doing, is beginning to strike most people as naive. (2017: 159)

The narrator then goes on to say that while most people still believe in the idea of a private life and fall into the habit of thinking in terms of intentions and individual causes of action, these are bound to die out. Ulrich is transformed by encounters with the probabilistic and statistical discourses of modern science, as he considers himself to be a random product of interacting forces over which he has no control. These forces are numerous and often mutually independent, which means that inventing a coherent story to strengthen the 'I' serves no purpose in the modern, complex and fluid world. Ulrich – in a dandyist spirit of thinking *à rebours* – flips the commonsensical system of epistemic coordinates on its head, claiming that it is not the 'I' which imposes its free will on the outside world but, on the contrary, it is the contingent network of circumstances that produces the 'I'.

> Have we not noticed that experiences have made themselves indepen-
> dent of people? (. . .) A world of qualities without a man has arisen, of
> experiences without the person who experiences them, and it almost
> looks as though ideally private experience is a thing of the past, and
> that the friendly burden of personal responsibility is to dissolve into a
> system of formulas of possible meanings. (158)

According to this view, a man is not a fixed and essential entity, but a collection of habits, thought patterns and roles to be played with others. Some of them complement each other, others are contradic-tory, so the final product, such as a person named Ulrich, is merely the result of interacting forces and systems of knowledge that stretch far beyond his control and perception.

Ulrich's personal attitude toward the impersonal discourses of modern science manifests itself throughout the novel, such as when he twists even the most personal topics of conversation so that his interlocutors find them uncomfortably abstract and inhuman. When discussing female beauty he speaks of 'the fatty tissue supporting the epidermis'; when broaching the topic of love, he responds with a long speech about 'the statistical curve that indicates the automatic rise and fall in the annual birthrate'; great metaphysical concepts, 'impossible to pin down' he sees as nothing but entropy (302) – a growing decay that comes with the advance of civilisation. In those lengthy daily lectures, thermodynamics and other statistical sciences, in which non-determinist notions like entropy, randomness and uncertainty play a key role, figure most prominently (see Kassung 2001). The choice of these particular sciences corresponds with Ulrich's anti-humanist (and anti-liberal) sentiments, because it was statistics that dealt the most important blow to the humanist paradigm centred on the idea of a free human will.

As early as the nineteenth century, many statisticians, or other scientists who relied on statistics in their work, were alarmed by the troubling consequences of applying statistical reasoning in the social sciences. A founding father of statistics, Adolphe Quetelet, 'took considerable pains to define the small domain of free will that surrounded each individual' (Gigerenzer et al. 1989: 44), yet still admitted that individual agency has a negligible effect on society as a whole. Quetelet was particularly drawn to the phenomenon of crime, which, in his view, should not be attributed to individual will but to 'the customs of that concrete being that we call the people, and that we regard as endowed with its own will and customs, from which it is difficult to make it depart' (1847) (in Gigerenzer: 43). Quetelet concludes from this that actions of free individuals become scientifically significant only when amassed to produce regularities visible in collected data through the lens of statistical analysis. He even stated that: 'It is society that prepares the crime; the guilty person is only the instrument who executes it. The victim on the scaffold is in a certain way the expiatory victim of society. His crime is the fruit of the circumstances in which he finds himself' (1832) (in Hacking 1990: 114). Another phenomenon that sparked the interest of many early sociologists and philosophers was suicide, which was long considered the ultimate act of freedom. The decision to one's own life was perceived as a sin, but also as a defiance of the natural order and the instinctual urge to live. However, when first large pools of data were collected, it turned out that both suicide and crime rates were relatively stable. This revelation suggested that what was previously considered an undisputable act of free will could be, in fact, socially determined. In other words, a violent crime or taking one's own life was no longer an individual decision, it was a statistical occurrence brought about by the society which literally produced the criminal.

Ulrich's views on the matter are in agreement with the doctrine of statistical objectivism when he declares the following:

> The scientific mind sees kindness only as a special form of egotism; brings emotions into line with glandular secretions; notes that eight or nine tenths of a human being consists of water; explains our celebrated moral freedom as an automatic mental by-product of free trade; reduces beauty to good digestion and the proper distribution of fatty tissue; graphs the annual statistical curves of births and suicides to show that our most intimate personal decisions are programmed behavior; sees a connection between ecstasy and mental disease; equates the anus and the mouth as the rectal and the oral openings at either end of the same tube – such ideas, which expose the trick, as it were, behind the magic of human illusions, can always count on a kind of prejudice in their favor as being impeccably scientific. (327–8)

Just like Duchamp's creator, Musil's individual is 'the interplay between inner and outer' (272), occurring on the border between personal background and impersonal determinations. For Ulrich, recognising the non-essential status of human identity is not just a matter of abstract considerations to be agreed upon during academic disputes and then forgotten in everyday life. It is, rather, a method of addressing oneself even in the most personal contexts and experiencing one's lack of solid ground aesthetically. He thinks of his own life in terms of a scientific experiment which demands he be as unbiased as possible: 'Comparing the world to a laboratory had rekindled an old idea in his mind. Formerly he had thought of the kind of life that would appeal to him as a vast experimental station for trying out the best ways of being a man and discovering new ones' (160). Naturally, he is not studying his life to acquire any sort of objective results. Rather, he is attempting to study it in the form of an essay – a quasi-scientific approach to investigating reality, described by Musil in his short text 'On the Essay' as a literary tool for producing probabilistic knowledge. The essay, he claims, is the strictest possible form of conducting research into those areas of existence where mathematical precision cannot be attained. Essayistic reasoning is never 'objective', but still puts important constraints on subjective expression: it is bound by the principle of rationality; it aims, one way or another, to find truth; it seeks to establish order; it moves forward by connecting thoughts (not characters); it presents evidence and sets out to investigate it. Nevertheless, it is not the same as logic or mathematics, as it touches on those aspects of reality which are, in principle, irreducible to simple systems of laws. Musil reiterates Maeterlinck in saying that the essay surpasses reductionist intentions of scientific procedure, making it possible to speculate on possible outcomes. He writes that the essay provides conjunctural knowledge in that it 'gives three good probabilities' '[i]nstead of a truth' (. . .). Then he adds that there are areas of life where 'it is not the truth that dominates, and in which probability is something more than a discussion about the status of probabilistic knowledge' (1990: 49), meaning, probably, those grey areas where contingency prevails over strict causality. Crucially, for Musil, this is not some mysterious and unknowable world, but a reality 'more encompassing and conceptually less pure', which is still graspable with intuitive thought. Here, thought cannot be reduced to operations on symbols: it is the process of organising (composing) mental phenomena of a varied, often imprecise nature, 'personal experiences, including mental ones . . . [p]reserved in a series of complexes interwoven with trains of thought' (1990: 50).

Ulrich is a prototypical essayist (*Möglichkeitsmensch*) who possesses what Musil calls 'a sense of possibility' (2017: 11). He is not interested in knowing the truth, but in making conjunctures, constantly questioning dominant narratives, offering alternative interpretations and hypothesising about what could happen. Instead of acting in his own self-interest, he tends to over-analyse things, moving through life in a state of perpetual, distracted daydreaming. As a possibilist, he avoids being determined by a set of qualities which would limit the possible scope of his behaviour. Thomas Harrison comments:

> So the sense of possibility could be defined outright as the ability to conceive of everything there might be just as well, and to attach no more importance to what is than to what is not. The consequences of so creative a disposition can be remarkable, and may, regrettably, often make what people admire seem wrong, and what is taboo permissible, or, also, make both a matter of indifference. (2014)

While rational and analytical, Ulrich still remains utterly indifferent to practical matters. Only the most improbable activities grab his attention: 'His extraordinary indifference to the life snapping at the bait is matched by the risk he runs of doing utterly eccentric things' (12). By replacing the notion of reality with possibility, Ulrich practises an experimental form of existence which can be best described as rational eccentricism. As he rejects discernible qualities which would define the reality of his identity, he also denounces any reified image of reality, especially one conceived through modern rationalism, claiming that the sole role of intellect is to be endlessly creating new knowledge, transforming reality in a process of unbound speculation. Through the notion of possibility borrowed from the probabilistic discourses of science, he establishes a link between modern rationalism and avant-garde creativity. In a chapter significantly titled 'If there is a sense of reality, there must also be a sense of possibility', Musil attaches special importance to this existential attitude:

> A possible experience or truth is not the same as an actual experience or truth minus its 'reality value' but has – according to its partisans, at least – something quite divine about it, a fire, a soaring, a readiness to build and a conscious utopianism that does not shrink from reality but sees it as a project, something yet to be invented. After all, the earth is not that old, and was apparently never so ready as now to give birth to its full potential.

In contrast to modern realists, a possibilist like Ulrich strives to reimagine reality, to challenge established norms and notions of what is real and what is not, continuously exploring the earth's possibilities. Confronted with the epistemological crisis of late modernity, on the one hand, and the danger of life's mechanisation in industrial society on the other, he wilfully and joyfully embraces a state of existential uprootedness, embodying the 'dissolution of the anthropocentric point of view' (Musil 2017: 159).

As evidenced by the example of Marcel Duchamp, Musil's invention, 'the man without qualities' who lives 'hypothetically' and experimentally for the sole purpose of challenging norms and probing the possible, was not a utopian (unattainable) ideal, but a mode of existence to which one could actually devote one's life. For both characters, whose lives transcended the duality of fact and fiction, chance and change were the most important values, almost synonymous with freedom, whereas the concept of possibility – understood as being just as important as reality – paved the way for them to devise novel modes of living. The main difference, of a very practical nature, between the fictional Ulrich and real Marcel lay, however, in their different habitus. If Musil's protagonist experienced a state of uprootedness spending his whole life in Vienna, free of economical obligations, Duchamp chose exile to escape France and all the securities provided to him by his homeland. By constantly shifting grounds, he managed to put Musil's idea of a groundless, probabilistic existence into practice, never settling on any one fixed identity.[6]

---

[6] A similar hero can be found in Tristan Tzara's statistically informed poem *Approximate Man*, whose protagonist also tries to come to terms with his lack of identity and the 'unspeakable punishments / of shock and incalculable weariness for no result / tormented as we are by microscopic predictions / poor souls unable to avert our gaze from the heel of death' (36).

Chapter 5

Cage: Music as Camera for Capturing Chance[1]

In an interview conducted in 1965 – which now reads more like the transcript of an interrogation – Michael Kirby and Richard Schechner asked John Cage for his definition of theatre. Right from the start, both interviewers cornered Cage, trying to make him restrain himself and make sense of his hazy concepts. In the mid-1960s, Cage, who became a very influential figure among the young experimenters and neo-avant-gardists in New York and across the Atlantic, was developing his philosophy of art to encompass the broadest possible range of creative activities, both human and non-human. His writings in that period were far more poetic and disorganised than before, lacking the precision he sought in the previous decade. However, under steady pressure from Kirby and Schechner, Cage finally disclosed that his approach toward artistic experiments was less relaxed and inclusive than was commonly assumed. In fact, his philosophy of art turned out to be quite restrictive. When Schechner mocked happeners for primarily making poorly rehearsed theatre, Cage rejoined: 'If there are intentions, then there should be every effort made to realize those intentions. Otherwise carelessness takes over' (Cage, Kirby and Schechner 1965: 56). He could have been an egalitarian democrat in the war against cultural elitism, yet his revolutionary methods were quite Jacobinian (see Piekut 2011: 35–40). Only a few sentences on, Cage proceeded to explain that he liked to avoid 'carelessness' by subjecting his performers to technological arrangements that were so deliberately complex and incomprehensible as to preclude intentional action. Schechner, obviously unnerved by the idea, commented

---

[1] This chapter was originally published as 'Capturing the World with Performance: John Cage's Probabilistic Aesthetics for the Digital Age' in *The Drama Review*, 2019; 63: 4 (244), 33–56.

that this had to imply 'us[ing] a machine which [. . .] short circuit[s] human intention' (57). Regrettably, the topic was dropped here, but it is significant that Cage was in no way troubled by this reading of his intentions.

Kirby and Schechner's investigation gives just an inkling of the relevance in Cage's 'performance theory' of his admiration for new technologies and the alienation they afford. Taking a step back and looking closely at his earlier, seminal writings from the 1950s, Cage's techno-enthusiastic world view is evident in the camera metaphor he used to describe musical composition. His theory of composition and indeterminate performance responded to the emergence of a new epistemology of information and conveyed a probabilistic agenda: a way of apprehending reality that only became possible after the first computers were up and running. As a performance theorist, Cage influenced generations of artists exploring matters such as spontaneity, emergence, liveness and indeterminacy. The idea of composition-as-camera does not simply imply that musical notation is a form of technology. There is far more depth to this metaphor, which also suggests a certain understanding of the materiality of sound, and corresponds well with Cage's epistemological views.

Obviously, these topics have already been raised separately and discussed in detail by numerous scholars. It might even seem that, given the vast wealth of 'Cage studies', any attempt at revisiting his legacy is a desperate or arrogant endeavour (for an overview of source literature see Joseph 2016: 8–11). Moreover, my investigations of Cage's aesthetic programme are prompted by an old question: Is it possible to explain his methodical and disciplined approach to indeterminacy and his admiration for the spontaneity of material processes within a coherent epistemological framework? For most scholars, like Benjamin Piekut, this task should fail, because of the fundamental contradiction at the heart of Cage's art: it constantly wavered between the (modernist) intention to gain 'unmediated access to the world' and the actual practice of artificially constructing ecologies 'of entities that mutually enhanced and defined one another through the event of the experiment' (2013: 135). My line of argument, devised to resolve this apparent paradox, is based on the assumption that the contradiction is not inherent in Cage's art but stems from its situation in art history. In my opinion, Cage's overt wariness of European modernism and conservative art institutions (see Silverman 2012: 150–1) – in which, for lack of a better alternative, he worked – did not come out of nowhere, but expressed a true sense of incongruity. Because Cage identified with the cybernetic utopianism of Buckminster Fuller, or Marshall McLuhan, he could

not feel really at home within the professionalised and compartmen-
talised reality of the modern world. I would even argue that he strug-
gled with the implications of being labelled a composer or an artist.
In fact, the assumption that Cage was an artist should be questioned
for the simple reason that his use of composition and live performance was not intended to celebrate human individuality, community or cultural heritage. Cage himself often admitted that he wanted
to celebrate the chance event, regardless of its origin. To this end the
boundary between inhuman matter and human perception had to
be abolished. Composition and performance could be regarded as
technologies that served to capture and transform concrete physical
space (of objects and relations observed from a fixed point) into an
abstract space of possibility (or shifting probabilities). As such, it is
not that, as Branden W. Joseph suggests, '[for] Cage, sound ontologi-
cally is a virtual multiplicity' (2016: 152). It is rather *made into* a
virtual multiplicity using the artificial frame of performance, just as
reality is turned into hyperreality using simulation technologies.

There is one more reason why I believe Cage's legacy is worth
revisiting. He is often depicted in performance studies textbooks as
a pivotal artist whose ideas and work established the foundations of
the discipline. Both Schechner and Fischer-Lichte have pointed out
that, beyond being an influential figure for many performance artists,
Cage was also the thinker who opened up the possibility of apprehending everyday life in terms of performance. Such statements are
certainly true, but require further elucidation, as Cage's understanding of performance as a probabilistic process does not easily fit the
conceptual frameworks of performance studies (Fischer-Lichte 2008:
19; Schechner 2013: 44). In my view, Cage's trademark aestheticisation of non-intentional expressivity of matter cannot be discussed
independent of his interests in information technologies and conjec-
tural sciences. In other words, his idea that every event can be experienced as an aesthetic performance harbours more complexity than is
generally assumed. Technical and scientific discourses informed and
decisively shaped his interests in the performativity of matter and the
'liveness' of sound – after all, the idea of the eternal persistence of
music occurred to him in the anechoic chamber at Harvard University, and not in a remote hermitage (Cage 1961: 9). As such, it was
not only a case of Cage using media to create a new sense of liveness,
which, many decades later, Philip Auslander found to be a prevalent
feature of many popular live performances (2011). It was exactly
the reverse: his interests in the durational, material and contingent
aspects of sound – the here and now of the performance space – were
motivated by his early involvements in new sound technologies and

experimental sciences. The camera as a metaphor for composition was thus not incidental, but essential to conceptualising performance in the technical age. For Cage, seeing the world as a performance was like putting on AR glasses – not coincidentally, now being produced by major photo companies (Fuijitsu, Nikon) – and perceiving the surrounding reality as a continuous flow of information.

The Camera: Making Indeterminacy

The camera, like most of Cage's metaphors, appeared in his texts time and again, yet it was elaborated on in detail only once, in the 'Indeterminacy' lecture, part of a larger series called *Composition as Process*. It was written in 1958, presented during the Darmstadt International Summer Courses, and published three years later in the United States. The text concerns – and Cage tests the reader's patience by making the point over and over again – 'composition which is indeterminate with respect to its performance' (1961: 35). In example after example, he discusses compositions by Morton Feldmann, Earle Brown, Christian Wolff, Karlheinz Stockhausen and himself, confirming their indeterminate character. The text is full of bold statements and provocative metaphors – more a manifesto than an analytical essay – yet Cage counters its 'pontifical' tone by making the letters so small that it soon becomes extremely painful and annoying to read. To make matters worse, Cage repeats the same phrases and references so often that one might wonder if he is actually trying to make a point. Given the annoyingly small font size, repetitive rhythm and formal style, 'Indeterminacy' reads more like a dry manual than an artistic essay. It is also worth noting that its exceptional tedium is more evocative of the factory hall than the monasterial garden. In today's terms, it might be considered an elaborate exercise in the art of trolling.

This frustratingly monotonous style is a good match for Cage's peculiar choice of metaphors. To begin with, he compares the ideal musician performing his *Music of Changes* (1951) to 'a contractor who, following an architect's blueprint, constructs a building' (36). Then, he describes the composition itself as 'an object more inhuman than human' with 'an alarming aspect of a Frankenstein monster' (who, unsurprisingly, is the only creature Cage borrows from the Romantic bestiary). Next on the list of industrial terms are 'light[s] in metropolitan thoroughfare control', used in the notes for Feldman's *Intersection 3* to describe the table, which was to be filled in by the performer before the show. Despite the obviously bland connotations

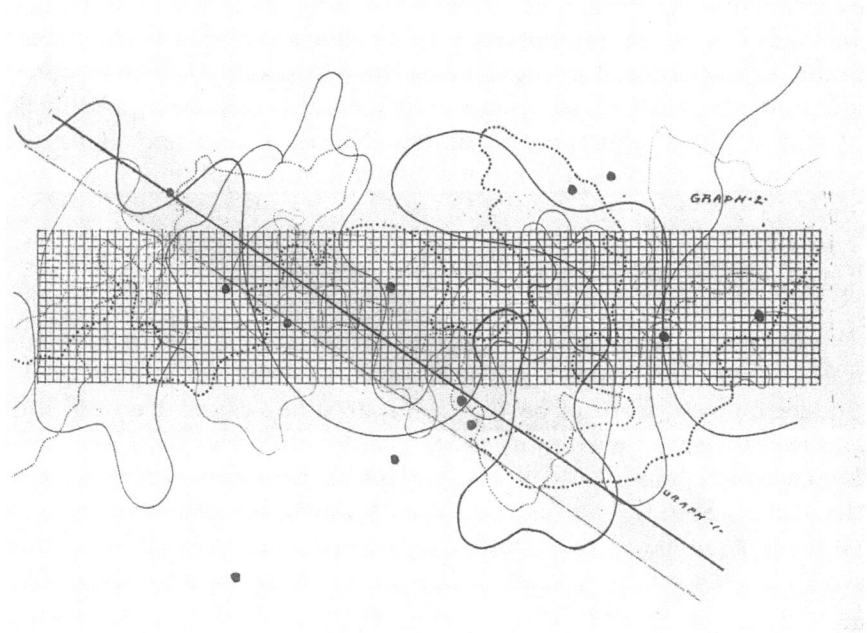

Figure 5.1 John Cage, Drawing for 'Fontana Mix' © Copyright 1958 by Henmar Press, Inc., New York. With kind permission of C. F. Peters Ltd & Co. KG, Leipzig

of traffic lights, Cage actually evokes this image to compliment Feldman for giving the performer freedom to 'play [. . .] sounds in the range indicated at any time during the duration of the box, just as when driving an automobile one may cross an intersection at any time during the green light' (36). There is something bizarre in Cage purposefully linking artistic freedom with the strict rules of traffic, and free action with submission to external control, especially if we take into account how strikingly different were beatnik associations between cars and freedom. Unlike the rebellious driver expressing himself through limitless exploration, such as Sal Paradise in Jack Kerouac's *On the Road*, who romanticises the automobile as a machine which makes 'the wind roar' and 'the plains unfold like a roll of paper' ([1957] 1976: 234), Cage's driver is stuck in traffic and can only determine the speed of his car or pick one of four directions. This image of freedom as a property of composition insinuates a limited number of outcomes within a confined space of probability: a virtual reality constructed to model behaviour of random processes. It obviously contradicts the common perspective of associating the artist's freedom with following intuition or intentions into uncharted territories, creating something new and exceptional. And Cage was a man of his word: as reported by James Pritchett, composing *Music of Changes* was both time-consuming and meticulous, as every single note had to be produced using several charts (1993: 78–88).

As if this were not enough, Cage's anti-humanist stance becomes even more apparent in an image that conjures the techno-enthusiasm of the Italian Futurists. Discussing Christian Wolff's *Duo II for Pianists*, Cage paints a fascinating picture of a man lost in the complexities of the modern world, which echoes Umberto Boccioni's famous triptych, *States of Mind*:

> The function of the performer in the case of *Duo II for Pianists* is comparable to that of a traveler who must constantly be catching trains the departures of which have not been [pre-]announced but which are in the process of being announced. He must be continually ready to go, alert to the situation, and responsible. (1961: 39)

This is an image of a person both overwhelmed and engrossed by the complexity of the surrounding infrastructure. Instead of criticising it, Cage actually praises this state of cognitive overstimulation. He even takes it as his task to strip the audience of their liberal 'encapsulation' which preserves individuality. Thus, the performers in Wolff's piece are not positioned as people who consciously follow predetermined instructions, and, depending on the composition, interpret the work to adhere to his aesthetic preferences, sensibility, education or intuition. After committing themselves to the performance, they must navigate the complexities of the composition and react to unpredictable stimuli from the environment. Freedom and creativity stand for the inability to control (and thus predict) action – even one's own – within the set boundaries of a technical system. In its disavowal of subjectivity, this sincere appraisal of dehumanisation through technology evokes radical transhumanist or accelerationist philosophies, shared in the 1960s only by niche groups of techno-nerds who fantasised a new, alienated humanity, such as John Brockman (2014), who, to this day, works as a publisher and editor of *Edge*, an online journal devoted to speculation about the cultural and epistemological significance of scientific breakthroughs and technological innovations.

It is safe to say that the 'dehumanizing' aspect of technological progress did not bother Cage in the slightest. With the notable exception of his lifelong disregard for the escapist pleasures of home listening (in Kostelanetz [1987] 2003: 88), Cage's stand on electronic media was never negative, even when it came to other forms of art or in daily life. Despite the post-war generation's growing tendency to denigrate technology and vilify the mechanisation of everyday existence, throughout the 1960s and until the very end of his life, Cage remained enthusiastic about the existential fruits of technological progress (see his last lecture, 'Overpopulation and Art' [in Perloff and Junkerman 1994: 14–38]). It is worth noting in this regard that even his poetic style of mesostics, which his writings employed to avoid clarity, relied on an unconditional submission to arbitrary rules of composition and additional randomisation procedures. His interest in artificially produced indeterminacy suggests that Cage was excited about technological development precisely because it pushed (old) humanity over the brink and into the realm of non-human unpredictability. He might have been worried about the looming possibility of overpopulation, but he thought the idea of 'robots making robots' was 'exhilarating' (Silverman 2012: 402–3). Many of his experimental performances – including the most iconic, like *Variations V* (1965), *Reunion* (1968) and *HPSCHD* (1969) – were filled with electronic devices connected in mysterious ways. The cables were tangled on the floor, making it almost impossible for the viewer to grasp the 'logic' behind the system, which was itself, at least to some extent, often designed by chance and relied on random events. These performances can be seen as semi-artificial microcosms that offered a glimpse into the posthuman future after technological singularity. Not incidentally, Cage's approach to experimental art favoured unpredictable results: 'An experimental action is one the outcome of which is not foreseen. Being unforeseen, this action is not concerned with its excuse' (1961: 39). Just as in a science lab, instruments bleeped, buzzed, crackled, glimmered, etc., but despite their obvious interconnection these audiovisual happenings made no sense to the human beings gathered around them. Commenting on another multimedia piece, *Time and Space Concepts in Music and Visual Art* (1978), Cage admitted that he saw nothing wrong with making non-human compositions: '[I]f I have the opportunity to continue working, I think the work will resemble more and more, not the work of a person, but something that might have happened, even if the person weren't there' (in Kostelanetz [1987] 2003: 52). Cage already had this radical idea in the 1950s when he spoke of composition as a 'nonhuman' device.

Taking all this into account, it seems plausible that in claiming that 'the function of the performer in the case of the *Intersection 3* is that of a photographer who on obtaining a camera uses it to take a picture' (1961: 36), Cage was not implying that composition should be regarded as a tool to be used for a purpose. The unfortunate appearance of the verb 'use', suggesting some form of purposeful conduct, is contradicted by an earlier statement pertaining to *Music of Changes*, which argues that a musical composition 'comes together to control a human being' (1961: 36). The juxtaposition of composition and camera is thus ambivalent – it signifies an object that is a tool in its user's hands, but also an 'inhuman' monster (body snatcher) that controls its servant. This apparent paradox can be solved by the philosophy of Vilém Flusser, who argued: 'The apparatus does as the photographer desires, but the photographer can only desire what the apparatus can do' ([1985] 2011: 20). Flusser's philosophy of media and communication, developed mostly in the 1970s and 1980s in Brazil, bears a striking resemblance to Cage's early intuitions about the existential importance of photography. It is even possible that Flusser 'borrowed' the camera metaphor from Cage, but legal issues of intellectual property are of lesser importance here.[2] I find it more interesting to compare Cage's train of thought with Flusser's, who concluded that photography lay at the heart of an unacknowledged cultural revolution that took humanity from a historical period shaped by the written word into a posthistorical age of information technologies.

Flusser saw the camera as the first human invention to produce images that were not intentionally created representations but technical 'projections', translating contingent matter into 'probabilistic' information. What made the photographic image so exceptional was the concealment of the creative process inside a black box: after the button was pressed and the shutter was released, the chemical process unravelled by itself. Taking pictures required no training in chemistry or digital image processing. In comparison, 'traditional' pictures were produced through conscious intentional action. It was impossible to make a painting devoid of human meaning. Even if based on an abstract set of rules, like the Renaissance perspective, these would still have to be internalised by the artist. Every painter perceived the world through their own eyes and represented it according to the prevailing cultural codes. In this sense, even the most 'realist' works were still the 'self-expressions' Cage so despised, because it was impossible for a human to paint what could not be perceived, imagined or invented, based on their own knowledge. Human art could only produce human objects.[3]

---

[2] It is only possible to assert with certainty that none of Cage's writings are to be found in Flusser's travel library (now in the Flusser-Archiv at Universität der Künste in Berlin).

[3] An interesting exception to his rule came in music: *Musikalische Würfelspiele* – musical dice games created to generate new pieces from precomposed sets of notes. These appeared in the mid-eighteenth century: in practice 'what constrained the choice of figures [. . .] were the claims of taste, coherent expression and propriety, given the genre of work being composed' (Meyer 1994: 193). Cage and Lejaren Hiller inserted such games into their famous *HPSCHD*, which premiered in 1969.

In the age of photography, the individual nervous system no longer needs to be in full control of the creative process. Of course, utterly non-human imagery – in the literal sense of 'nonhuman' – has only recently become widespread, thanks to drones, satellites, CCTV and so forth (see Zylinska 2017). However, before the world was blessed with these technologies, Flusser already saw that the non-human aspect of photography had been inscribed in its nature. The camera – itself an invention of rational thought and experimentation – captures the world without imposing causality or intentionality upon it. It records and stores information 'about' the world in codes that make no sense to humans. For instance, a photo captured on celluloid is a coded message which has to be chemically decoded in a darkroom. This process can be considered 'inhuman', because it occurs on a microscopic level that is completely meaningless to *Homo sapiens*. The same logic also applies for digital photography: binary code is not only extremely difficult for humans to master, but also would be very inefficient in everyday use, given the physical properties of our vocal apparatus. Even programmers do not use 0s and 1s; they depend on compilers that automatically translate code written in Roman alphabet and Arabic numerals into lower level language understood by the machine. Switching from alphanumerical to digital is thus highly improbable unless humanity alters its biological 'wetware'. Another important property of the camera, with regard to its inhuman character, is that it does not have to understand the world to take a picture (although new digital devices equipped with neural networks to recognise patterns can possess a non-human understanding of it); it produces 'mosaics assembled from particles' (Flusser 2011: 6) by the 'capturing and holding of approaching particles or waves from the environment' (42). Even if we still tend to assign meanings to photographs:

> What remains are particles without dimension that can be neither grasped nor represented nor understood. They are inaccessible to hands, eyes, or fingers. But they can be calculated (calculus, 'pebbles') and can, by means of special apparatuses equipped with keys, be computed. The gesture of tapping with the fingertips on the keys of an apparatus can be called 'calculate and compute.' It makes mosaic-like combinations of particles possible, technical images, a computed universe in which particles are assembled into visible images. (10)

In other words, technical images offer us a new and unseen perspective on reality: de-signified, meaningless, yet orderly and perfectly rational. Even if, on the surface, photos seem like representations – and far more detailed than ever imagined – they are also instruments that remove intentionality from art and human culture. The same reasoning applies to Cage's 'technical compositions': they capture a material reality within a rigid frame of instructions that remove human intentions. *4'33"* (1952) is a perfect example. Its photographic and inhuman aspects constrain the performer's meaningful and intentional behaviour, exposing the contingent and performative side of sonic phenomena. Curtailing individual expression releases the non-human expressivity of matter, refracting light or emitting sounds to be captured. As a type of camera, *4'33"* is a technical device that transforms the concert hall into a multidimensional space of possibility.[4] Actual events determine the probability (virtuality) of the ones that follow (see also Temperley 2007).

Cage's Self-Determination

Another equally important reason to sacrifice the performer's personal freedom arose from Cage's conviction that intentional behaviour was just one form of self-determination. In outlining different ways of executing indeterminate compositions, he enumerated six basic strategies: (1) 'do[ing] this in an organized way which may be subjected successfully to analysis'; (2) 'feel[ing] his way, following the dictates of his ego'; (3) acting 'more or less unknowingly [. . .] as in automatic writing'; (4) following 'collective [. . .] inclinations of the species'; (5) entering the state of 'deep sleep [. . .] identifying there with no matter what eventuality'; (6) and finally 'perform[ing] his function of photographer arbitrarily [. . .] by employing some operation exterior to his mind: tables of random numbers, following the scientific interest in probability; or chance operations' (1961: 36–7). Though Cage saw no essential difference between these strategies, he preferred the less egotistic, as they could yield more unpredictable results. His condemnation of 'ego' in art was motivated by probabilistic reason: following the dictates of the ego determines behaviour and narrows the field of possibility for producing sounds, as it does in making any choice. This is only one programme that allowed him to turn possibility into necessity. Moreover, it had been operating in Western culture for so long and in so many minds that there was no reason not to replace it (or update it, at least). From Cage's perspective on art, which he sees as the exploration of unpredictability, intentional behaviour offers no surplus value. Holding

---

[4] Cage uses this exact notion in his later text, 'The Future of Music': 'First of all: the activities of many composers, particularly Feldman and Wolff, who have made their compositions indeterminate, so that performers, rather than merely doing what they are told to do, have the opportunity to use their own faculties, to make decisions in a field of possibilities, to cooperate, that is, in a particular musical undertaking' (1981: 181).

a camera, the performer can only choose from among different paths of self-determination in a field of possibilities predetermined by the device. The only 'choice' left is from different ways of constricting the performer's freedom. Furthermore, if one assumes that there is no substantial difference between human and automated creativity, then there is also no reason why we ought not to use the I Ching or scientific 'tables of random numbers' in composing, which also provide a predetermined set of possibilities (for *Études Australes* [1974–5] Cage both consulted the oracle and left blank spaces for the performer to fill in [in Kostelanetz (1987) 2003: 86–8]). Cage even suggested that the process of making art was, to some extent, automatic, citing Feldman's remark about feeling dead while composing (1961: 37). Apparently, making an indeterminate performance involves a lot of heavy equipment, discipline and dead characters. As in the old stories of people terrified that the camera would steal their souls, for Cage it also plays the part of a deadly instrument of dehumanisation. Only his attitude toward this process is strikingly different: he very actively pursued the goal of losing his soul.

Cage came in for a lot of criticism over this ambiance of 'deadness' and of 'silencing' human expression, not only from those, like the composer Eric Salzman, panicking that 'the denial of will, of intelligence, of consciousness can only lead to spiritual nothingness and death' (1960: X12), but even from cautious academics. Most of them saw the idea of silencing the individual as tantamount to an attack on human freedom. Timothy Morton, for instance, accused Cage of harbouring a postpolitical agenda that anticipated the rise of the neoliberal pseudo-utopia:

> The musical perceptions of John Cage, celebrating a notion of quietness, evoke a communitarian suburban or libertarian form of quiet that is also static in a political sense – there is no chance of progress, just an endless application of laws. Quiet is a meaningful, continuous absence of noise, often with strict legal definitions. (2007: 102)

We might take two main issues with this critique (also iterated by other scholars [see Michaels 2011]). First of all, it rests on the assumption that Cage's indeterminate art is actually 'quiet' and 'static'. Obviously, this claim does not apply to most of his live performances, many of which were noisy, disturbing and hectic. Precisely for this reason, Cage never really entered the canon of the middle class's favorite composers. As cleverly noted by Slavoj Žižek, the sublime and turbulent music of Ludwig Beethoven still performs much better than Cage's silent pieces in the role of an ideologically transparent filler (2014: 70). A report prepared in 2013 by the League of American Orchestras only confirms this intuition: Beethoven ranks first among the most played composers, whereas Cage did not even make it to the top 10 on the national list (LoAO 2014). Secondly, Morton's argument ignores the anti-individualistic aspect of Cage's music, which favours deindividuation – a loss, rather than a strengthening of self through disruptive technology. If there is a grain of truth in Morton's argument, it is the problematic nature of Cage's democratic fervour. In Cagean techno-enthusiastic epistemology, the notion of freedom, which post-Judeo-Christian anthropologies associate with personal will, draws from the indeterminacy of a larger system. The I Ching, which assumes the role of an inhuman partner in the conversation on composition (in Kostelanetz [1987] 2003: 87), disrupts the intimate bond between artist and his work, just as 'randomisation' in the natural sciences removes the slant of the individual experimenter.

From this standpoint, Cage's aesthetic strategies of depersonalisation emerge as self-conscious attempts at reprogramming habitual methods of creativity, and redefining music to fit the new epistemological horizon established by the probabilistic sciences. A bridge between indeterminate composition and these new sciences was established by American musician and philosopher Leonard B. Meyer, who was associated with the University of Chicago in the latter half of the 1950s. Even though it is difficult to determine if Cage was aware of Meyer's work while writing 'Indeterminacy', it is interesting to compare Meyer's theory to Cage's, as it introduced a perspective on music similarly informed by cybernetics and the information sciences. In his groundbreaking exploration of the perception of music, Meyer proposed approaching every piece of music as a type of Markov process – that is, a sequence of events in which each state's probability depends on the one previous. As the piece unfolds before the audience, and 'the probability of a particular conclusion increases, uncertainty, information, and meaning will necessarily decrease' (1957: 419). In Meyer's view, music is a code that we unravel probabilistically; its expression relies on building expectations and violating them. Listening, in turn, is an act of inference: after hearing the first sounds of a composition, the listener begins looking for regularities (redundancies), which can then be used to infer what is to come in the piece. The ability to find patterns allows the listener to (unconsciously) create hypotheses about the underlying structure of the piece: its key, melody, tonality, metre, etc. Moreover, this process takes place within another probabilistic context – the culture:

> [M]usic may be meaningful in the sense that within the context of a particular musical style one tone or group of tones indicates – leads the practiced listener to expect – that another tone or group of tones will be forthcoming at some more or less specified point in the musical continuum. (413)

Meyer suggests that what we usually call 'personal taste' in music is, actually, an 'internalized probability system' that produces unconscious expectations. Although Meyer's essays of the 1960s were sceptical about Cage's musical choices, his probabilistic understanding of music gave him the intellectual tools needed to recognise that Cage was a very serious intellectual, whose music 'express[ed] a consistent, clearly defined set of attitudes and beliefs about nature's universe and man's place in it [. . .]' ([1967] 1994: 71) In other words, even if Meyer was convinced that listening to indeterminate music was weird, he recognised that it had political and existential significance. Only from the probabilistic point of view could the demand to replace (liberal) autonomy of choice make any sense as a new kind of freedom. In Cage's posthuman aesthetics, art was not bound by market pressures and social conventions; it was affirmed as a special social environment in which people celebrated nonliberal freedom in its joyous absurdity. By releasing the creative potential of inhuman agencies – 'an incalculable infinity of causes and effects' (Cage 1961: 47) – indeterminate performance aestheticised the absurdity of the new world in which the *anthropos*'s point of view was no longer unique or central. In sum, Cage takes Meyer's scientific theory (description) and turns it into art (practice).

Into the Age of Programs (and Out of Humanity)

Naturally, Cage's meaningless sound performances praising the impending posthuman age did not resonate well with liberal arts institutions. As the camera-composition was meant to provide a transparent (non-intentional and disinterested) frame for the contingent performance of matter within it, indeterminate pieces were better suited for a very different kind of space: the post-industrial warehouses where illegal raves were held in the early 1990s. Cage, who is rarely mentioned among the forefathers of rave culture, probably due to his quiet persona, should be considered an early prophet of this musical revolution. Simon Reynolds, who wrote extensively about rave culture, contended that:

> [T]echno is closer to the plastic arts or architecture than literature, in that it involves the creation of an imaginary environment. [. . .] Devoid of text, dance music and ambient are better understood using metaphors from the visual arts: 'the soundscape,' 'aural decor,' 'soundtrack for an imaginary movie,' 'audio-sculpture.' [. . .] From the text-biased vantage of rock criticism, dance music is troubling precisely because it seems to be all materiality and no meaning. Entirely an appeal to the body and the senses, it offers no food for thought. (2010: 51)

Rave experience is predominantly spatial, not only in the sense of completely losing track of time, but also in its appreciation of the physical environment (or translation of music into spatial events through hallucinogens). Moreover, the 'bedrock' of techno music – uncanny sounds of unknown origin, a rhythmic structure framing a communal experience, chemical enhancement of the experience, and hi-tech acoustics – gives freedom to the producers and DJs, who often create soundscapes which, under normal circumstances, might be considered unpleasant. But for neurochemically reprogrammed nervous systems, dissonant and abstract noises sampled from industrial environments may evoke pleasurable sensations. Furthermore, like Cage's compositions, techno music produces artificial, non-human structures filled with indeterminate and meaningless physical actions. Ravers do not contemplate music at a distance, they immerse themselves in a chaotic whirlpool held together by alienated sounds and pounding rhythms which, aided by the right substances, 'temporarily quieten the neurotic self, freeing the individual from anxiety and fear' (Reynolds 2010: 83). In many respects, Cage's cybernetic aesthetics in 'Indeterminacy' would resonate better in club culture than in conservative high art institutions. Unsurprisingly, it is not uncommon to find a monk meditating on the crowded dance floor of a techno club.[5]

I find the aforementioned misconceptions especially surprising given Cage's reluctance to call himself a composer (in Kostelanetz [1987] 2003: 111). It would be more accurate to call him a programmer, or sound engineer, even though his actual programming skills were non-existent. Frances White, a sound engineer who worked with Cage, recalled, 'He never seemed to have an interest in a hands-on approach to the equipment' (in Bernstein and Hatch 2001: 199). Being a technician in spirit, what distinguished Cage from professionals was his non-practical (purposeless) approach to technical media. Coming of age at the dawn of the information society, he rejected the old cultural values, hierarchies and institutions, and identified with a futuristic, depersonalised posthuman image of 'global society' altered 'through electronics so that world will go round by means of united intelligence rather than by means of divisive intelligence (politics [. . .])' (Cage 1967: 17). Instead of utilising new tools for the old culture's old purposes, he identified with the purposelessness of the new age of 'programs' based on a non-human system of governance. To

---

[5] In 1937 Cage predicted that 'the use of noise to make music will continue and increase until we reach a music produced through the aid of electrical instruments' (1961: 3).

explain this, it is useful to reintroduce Flusser, who argued that new computing machines, inhuman algorithms and conjunctural sciences ushered in a new way of making sense of the world. The 'programmatic' perspective, as he called this new paradigm, took the place of the 'finalistic' and 'causal' images that had dominated Western culture until the invention of the computer. Within a finalistic image, typically found in Judeo-Christian religions, reality and individual human existence were explained in terms of a linear narrative: monotheisms made purpose an immanent component of reality, framing the existence both of individuals and humanity as such – as stories. In turn, modern sciences replaced teleology with the universal rule of cause and effect. Here, human existence was caught in a complex net of causal relations. Importantly, these stories did not contradict each other, because causality does not rule out teleology. The existence of God in the modern world was not necessary to secure the illusion of grand coherence (this is the role of the Newtonian commandments), yet was a viable addition. Things became more complicated in the twentieth century, in which probability theory, statistics, quantum physics and chaos theory challenged the dogma of coherence. The programmatic perspective built on these foundations is irreconcilable with previous points of view, because it allows virtually anything to be explained as a program. The Bible, children's books, laws, local customs, Renaissance paintings, etc., can be written automatically by specialised programs and, at the same time, they are programs themselves, used for centuries to operate on human populations. The digital revolution revealed that programs are everywhere; they can be simply defined as:

> systems in which chance becomes necessity. They are games in which every virtuality, even the least probable, will be realized of necessity if the game is played for a sufficiently long time. [. . .] Absurdly improbable structures, such as the human brain, emerge necessarily in the course of the evolution of the program contained in genetic information, even though they had been entirely unpredictable in the amoeba. And they emerge by chance at a particular moment. Wonderful artworks such as *The Marriage of Figaro* emerge necessarily in the course of the evolution of the program contained in the initial project of Western culture. Although it is absurd to look for them in that initial project, for example, in Ancient Greek music. This is because, although they become necessary, such realizations emerge by chance in the course of the game. (Flusser 2013: 22)

The same rule applies to compositions like *4'33"* or *Imaginary Landscape No. 1* (1939), though the conditions for them to emerge could only be grasped after the Industrial Revolution and with the Digital Revolution in sight. Their relative unpredictability – from the perspective of Renaissance art, for instance – stems from the complexity of human culture, which evolves probabilistically (not by absolute chance, but also not deterministically) through the inter-action of semi-independent knowledge systems (discourses). As this process began accelerating in the eighteenth century, and new sets of rules and playing dice were added to the program of Western culture, it became necessary to accommodate such notions as complexity, randomness and probability into its operative logic. The unques-tionable success of new theories and devices built with probabilistic notions in mind came to contest the universality of the old program with its basic notions of nature, identity and substance. Cage's alien-ated and purposeless art is symptomatic of this new epistemological order. Its depersonalised aesthetic identity adheres to the logic of digital machines. In 'Indeterminacy', Cage approaches composition in purely functional terms: it is just another set of instructions for anonymous performers. It should be added that these instructions are executed in spaces (the concert hall, theatre) that also reflect a culture's rules and protocols. The shape of a building, the number of limbs on a performer, or the concept of a reasonable and responsible citizen might not be completely arbitrary, but at the same time, all are subject to alteration, sometimes by pure chance. As 'human pro-gramming is itself increasingly programmed by apparatus' (Flusser 2013: 25), culture becomes absurd, but at the same time, surprisingly liberated. Flusser was convinced that programmatic images allow us to formulate the problem of human freedom after it has been buried by deterministic sciences. As reality becomes reinvented in terms of programs, games and information, chance becomes fundamental to freedom. For the first time in human history, it made sense to pro-voke chance events – to contain them in scientific experiments and harness them for new information (see Hacking 1990), or simply to celebrate them and create an egalitarian community around the unforeseen. Cage surrendered to the latter vision.

Although Cage's relationship with information technologies has been debated on many occasions, no one, to the best of my knowledge, has acknowledged the link between the notion of inde-terminate performance and probabilistic information epistemolo-gies. For example, in *Records Ruin the Landscape* (2014), David Grubbs discusses Cage's experiments with sound reproduction and even includes a lengthy comment on the camera metaphor, but does not situate these ideas within a wider probabilistic context. While noting the abstract and depersonalising aspects of Cage's art, he fails to recognise that they belong to a different epistemology. For this reason, despite Grubbs's awareness of the polyvalent meaning of the camera analogy, and that photography here should not be 'understood to suggest creating a representational image' (100), Grubbs finds inconsistencies and paradoxes in what seems to me Cage's perfectly spotless (and soulless) aesthetic project. My point is, even if Cage's 'innovations that involved the recording studio and recorded sound tended to be specific to individual works' (104), they were still grounded in a coherent world view. It may initially seem contradictory to use sound recordings in making music and during live performances, and then deprecate listening to records at home. However, it all makes sense when we appreciate that Cage saw no essential difference between live and recorded sound. His disdain for recorded music (in Kostelanetz [1987] 2003: 118) had nothing to do with authenticity. He simply did not care for listening to records at home, which he saw as a predictable and solitary experience. Claiming otherwise would imply believing in the value of the message. The first piece in which Cage experimented with alternative methods of organising sound, *Imaginary Landscape No. 1* (1939), was meant to be broadcast on the radio. This is no coincidence, because the very idea of generating random sounds through compositions is unthinkable without recording technology. Giving voice to formless matter assumes that it is capable of transmission and reception, which already implies both sensitivity and comprehension.

Of course, there is a caveat. By 'comprehension' I mean only a very simple mechanical awareness that allows sound waves to be translated into electric signals (information). This inhuman awareness can be described in terms of Zen Buddhism – as Cage himself and many commentators have done – but it can also be interpreted as an unconditional identification with external information technologies, which is more congruent with the camera metaphor. More importantly, interpreting Cage's art through spiritual inspiration does not explain how he persuaded his contemporaries that impersonal work could be considered aesthetic in the first place. Replacing contemplation with meditation as a default mode for aesthetic experience presupposed an already transforming cultural environment. This new cultural milieu was aptly described by Friedrich Kittler, who argued that Western culture undermined its core values through technological progress, in particular with the invention of film and the gramophone (1999). These devices of 'technological reproduction' captured and stored data in an unprecedented way: with no need for a symbolic language to be rendered by trained specialists (like musicians). Cameras and phonographs provided the means to record, save and communicate the materiality of the physical world for the first time in the history of civilisation: 'Ever since that epochal change we have been in possession of storage technologies that can record and reproduce the very time flow of acoustic and optical data' (Kittler 1999: 3). The phonograph, which originally recorded sounds onto a tinfoil sheet wrapped around a rotating cylinder, picked up and stored a wide range of sonic events, both intentional (wanted) and random (unwanted). Importantly, these sounds were not stored as notes or as any other symbolic code humans alone understood, but as groove modulations barely perceptible to our sense of touch. Before the invention of the phonograph, sound and speech could only be represented as conventional symbols separated by standardised intervals. Music – especially in its classical incarnations – had to 'pass through the bottleneck of the signifier' (4), which could only index a very limited spectrum (field of possibility) of sounds. Classical notation had no room for hisses, twangs, whirrs, clicks, knocks or bleeps; not to mention accidental and/or ambient sounds. These were excluded from the realm of cultural production for practical reasons – a symbolic language representing the entirety of possible sounds would be too complex for any human being to master. The phonograph bypassed this problem, embracing (almost) all sounds, recording and replaying them in the indefinite future. Thus, noise took its place in human culture, simply because it could be transmitted from one generation to the next. The only obstacle to the widespread appropriation of noises into cultural production was human sensibility, which was accustomed to the limited possibilities of traditional notation. Eventually, as the ears and minds of the public slowly adapted to recorded noises, artists and bands like John Cage, Pierre Schaeffer, Throbbing Gristle and Nine Inch Nails could attempt to sell thwacks, whizzes and bangs as progressive aesthetics of subversion. However, by appropriating long-denigrated noises as art and culture, they only followed in the footsteps of Thomas Edison, Emile Berliner and Alexander Graham Bell, who created equipment that recorded every sound without prejudice.

Another side effect of this process was a challenge to the well-worn opposition between reality and representation, since this duality applied only to symbolic modes of communication. Although Cage was certainly not the first avant-garde composer to aestheticise noise, he was probably the first to try to reinvent the very definition of music to fit the transformed techno-cultural landscape. As I see it, if we envision performing as taking a picture through composition, it implies a complete refashioning of our aesthetic sensibility: it means framing spatio-temporal continuity to remove meaning and a sense of familiarity with material processes. On the one hand, it reduces the symbolic value of a performance (to absolute zero); on the other, it increases the informational value of the aesthetic (sensory) experience, which cannot be focused solely on the succession of sounds as notes (determined by the linearity of written notation). Ideally, an indeterminate musical performance should sensitise the listener to a wider range of sounds and the relations between them. Even if aiming for indeterminacy in art meant adopting a machinic perspective on reality, it also aimed to provide a positive and transformational experience for the listener. As Cage firmly stated at the end of his 1974 lecture, 'The Future of Music': 'The change is not disruptive. It is cheerful' (1981: 187).

Unexpected Pleasures

Finding pleasure in the unexpected was pivotal to Cage's aesthetics. Without an ear attuned to random 'interpenetrations' of sounds, attending an indeterminate performance is about as interesting as watching paint dry. The principal condition for enjoying indeterminacy is believing improbability (of connection) to be a value in itself. Then, and only then, can discovering unusual associations between seemingly unrelated phenomena be rewarding. Without assuming a probabilistic attitude, it is hard to imagine why anyone would want to endure the endless frustration of indeterminate art. From a modernist point of view, consensual participation in an indeterminate performance is ludicrous. Of course, we could argue that, despite his anti-elitist claims, Cage was at heart a modernist, because if we define art as the production of useless objects in autonomous space, then 4'33" is a perfect embodiment (see Rancière 2004). This means the audience's task remains the same: they enter the room, sit quietly until the end, and leave. On the surface, Cage remained faithful to the old art institutions throughout his life, while adjusting its rules and stretching its limits. However, things get more interesting when we leave behind these abstract considerations and zoom in on the situation of the individual in a performance space with nothing of interest to listen to. As it is impossible to infer the intention behind an indeterminate piece, even silent, introspective criticism – a distraction to stimulate the mind – becomes futile. To enjoy Cage's indeterminate pieces, one is compelled to become aware of sounds in their concreteness and to 'celebrate' them for their unexpectedness. What is remarkable about this quasi-meditative state of mind is that it depends on both internal discipline and the environment. To paraphrase Cage's famous statement, both serve the same purpose of 'purposelessness', that is, of designification of the phenomenal world (1961: 12). However, stripping the perceived reality from meanings, customarily embedded in things or events, should not be understood in terms of revealing the unmediated world. On the contrary, Cage's empty frames (of reference) substitute meanings with information. It can thus be argued that Cage's grand scheme to aestheticise machinic codes and probabilistic cognition through inhuman art serves to create this peculiar environment, which allows external reality to be experienced as 'empty'. In other words, reality becomes disenchanted rather than enchanted (as claimed by Fischer-Lichte [2008: 180]): art as information technology breaks the spell of the symbolic. This might be compared to the alleged effects of Vipassanā meditation, which raises awareness by observing sensations without imposing meaning on them, yet ultimately rests on individual experience. In turn, Cage's indeterminate performances create artificial space in which it is impossible to cling to any meaning. In these temporary shrines of randomness, people can gather and communicate, not only with each other, but also with technological devices, using the basic language of probabilistic information. The performance space becomes a micro-version of the extended mind spanning the globe electronically. When Cage was asked about the impact of computers on the concept of the 'audience', he predicted that they would eventually transform musical culture:

> I think that not the computer itself, but the ideas that I and others have, in the field of music, have done this already, so that we think of the concert more and more not as something that begins and ends but as a process that continues and sometimes [is] very long so that people could come and go – again, flexibility. And one of the things that's so annoying about concerts to people who are themselves, let us say, not music lovers, is this business of sitting in rows in a theater situation, and our tendency now is to remove that by having a space in which people can move or sit, go out, come in [. . .] in the course of the performance of music. (In Kostelanetz [1987] 2003: 115)

This statement proves that Cage strongly believed in the potential of information technologies to change the culture – even though using computers to advance social freedom ran counter to common sense. In the late 1960s, computers were seldom seen as tools for democracy. For most people, they were devices 'of dehumanization, of centralized bureaucracy and the rationalization of social life' (Turner 2006: 2). Most computers in that era were at least the size of a wardrobe, and were only accessible to highly trained professionals in research facilities. Despite this, and even though Cage was unable to establish a causal link between electronic music and a flexible crowd, he intuitively assumed that digital technologies would eventually transform how people organised themselves to experience music. To make sense of this, one has to assume that behind this weird intuition lay a utopian image of live electronic music as a platform for intercultural communication in which a celebration of performative indeterminacy played a crucial role. Branden W. Joseph points out that Cage's aesthetics – even in the 1950s – expressed a radical political project akin to that of his close friend Buckminster Fuller, who speculated that in the future technology would establish a 'uniform and interchangeable space' for a global society. Given the fact that Cage shared Fuller's techno-enthusiasm as well as his holistic mysticism, Cage's vision replaced uniformity with 'multiplicitous heterogeneity' (Joseph 2016: 95). In my view, this aesthetic utopia also had an important probabilistic dimension: to make it possible and avoid cultural conflicts, indeterminate performance implies that everyone perceives the world in terms of shifting probabilities; despite their differences, everyone absorbs incoming data to make predictions (by contrast, not everyone thinks logically, speaks Spanish or finds pleasure in melody). The whole idea of indeterminate performance, in which surprise is celebrated in and of itself, hinges on this (non)ideological assumption.

I find it productive to examine Cage's intuitions in light of recent cognitive theory (which shares many similarities with Meyer's observations), which asserts that perception is first and foremost a probabilistic process. According to Andy Clark, who translates new experimental research into a coherent theory of the mind, all cognitive processes are essentially predictive:

> To perceive the world is to meet the sensory signal with an apt stream of multilevel predictions. Those predictions aim to construct the incoming sensory signal 'from the top down' using stored knowledge about interacting distal causes. To accommodate the incoming sensory signal is already to understand quite a lot about the world. [. . .] An animal, or machine, that has that kind of grip on its world is already deep into the business of understanding the world. (2016: 6)

Every mental activity involves anticipating what will happen, verifying a hypothesis against sensory data, and then guessing again. Seeing or hearing are not passive data-receiving processes, they are always coupled with predictions. Even the most mundane tasks – finding and preparing food, going through a door without banging one's head, or driving a car – require an ability to process incoming data and predict changes in the immediate environment. Abstract thinking or daydreaming operates on the same logic, though here the brain relies more on itself to verify its hypotheses. From this perspective, durational arts like music or theatre can be seen as guessing games: the audience identifies patterns and infers how the action or melody will develop. From time to time, they must be surprised and forced to update their prediction model. If perceptiveness keeps you alive on the street, then erudition keeps you awake in postdramatic theatre. No matter where you are – on the street or in the theatre – a failure to predict (where the brick lands, how the pun ends) results in a loss – of life or of interest. While humans generally assume that they perceive a reality made up of interacting objects and apprehend the relationship between the mind and the world in terms of symbolic representations, it turns out that they neither see nor apprehend the world through symbols. Under the surface of perception and conscious thought lie 'encodings of complex intertwined distributions of probabilities, including estimations of our own sensory uncertainty' (Clark 2016: 168). Symbolic processing, logical thinking and even Western individuality are not, therefore, natural conditions; they are cultural programs running on probabilistic machines made out of living tissue. If anything connects all sentient beings on Earth, it is their fondness for gambling.

What unites most gamblers is that they generally prefer to win. In Clark's terms, we can say that most organisms strive to minimise 'prediction error' (26). This basically means that they struggle to minimise uncertainty, either by adjusting perception, or through action. When you hear an alarming sound on the street at night, you turn your head to see if it was the wind, a kitten or a zombie (if you recently watched *Night of the Living Dead*). In this particular case, your sense of sight allows you to select one of many 'assumptions' reeling in your mind; these are assigned different probabilities, depending on the context. If the noise was only a cat, the brain returns to its former task: balancing the body as it walks down the street, scanning for potential threats, and so on. If it turns out to be a zombie, the brain receives a new set of tasks: using the knowledge supplied by popular culture, it predicts the zombie's behaviour, assesses the risk of putting up a fight, finds an optimal escape route and begins updating the belief system in which zombies do not exist (rethinking the existential and political consequences can be postponed). In both cases, reducing uncertainty depends on thought and action. It can also rely on external devices, like mobile phones, which can be used to get your bearings in an unfamiliar environment, or to call for help. Predictive processing stretches beyond the limits of the human skull.

However, in many complex situations, like making or experiencing art, the rule of prediction, of error minimisation, which implies that organisms should ultimately seek safe surroundings, is limited. To be sure, even in normal conditions of everyday survival, the unrestricted pursuit of certainty would be counterproductive, resulting in self-destructive escapism. Avoiding error at all costs would lead organisms to search for an equivalent of a 'sensory deprivation chamber', where, without incoming data to disprove their hypotheses, they would enjoy a state of predictive perfection. In reality, biological organisms regularly break the rule of error minimisation and sometimes even seek novelties that contradict their assumptions. Scarcity of resources, external threats or simple boredom (too much danger or too much safety) usually drive animals to explore uncharted territories or try out new behaviours. Humans have even professionalised the art of 'making mistakes': art, science, philosophy and astrology are all 'examples of domains structured to support and encourage just such open-ended forms of exploration and novelty-seeking' (Clark 2016: 267). So even though culture in general can be considered a communication ecosystem that minimises uncertainty (see Rappaport 1971; Hutchins 2014) – because it establishes a shared system of symbols, values and patterns of behaviour – it is also filled with intentional 'gaps', in which unpredictability is nurtured and celebrated. Clearly, this explanation of the purpose of art (as unbound experimentation for its own sake) would have been very dear to John Cage.

There is another important reason to celebrate errors. In light of 'predictive processing' theory, the brain derives more value from the error than from the correct conclusion. When sensory signals do not match predictions, the differences are passed on to deeper layers of the brain's network, producing new predictions. 'It is the deviations from what is predicted that then carry the "news," quantified as difference ("the prediction error") between the actual current signal and the predicted one' (Clark 2016: 26). For the brain, it is the incorrect conclusion that counts as valuable information. Perception and cognition thus seem to run on mistakes that transmit new information. This relatively new idea confirms an old mathematical intuition expressed by Claude Shannon in the 1940s and probably known to Cage. Shannon worked on noise reduction in communication, coming up with a theoretical model of communication and the first definition of information. As stated by Warren Weaver, who provided a commentary on Shannon's mathematical text, the amount of information in a message is proportional to the 'freedom of choice in selecting the message' and is 'measured as a logarithm of the number of available choices' (1964: 9). Information is therefore a quantitative and probabilistic measure of freedom in communication. Its value rests on three main factors: the length of the message, the complexity of its code and its unpredictability. Contrary to common sense and all theories at the time, Shannon and Weaver suggested – partly from a philosophical standpoint – that information equals entropy. As such, its value should increase in tandem with the decrease of the 'orderliness' of a message. A long, difficult stream of babble, like *Finnegans Wake*, one of Cage's favourite books, carries a lot of information, because it is exceptionally difficult to predict what will happen next, even the next word in a sentence. By contrast, classical and harmonious works of art hold less information, as they contain a lot of redundancy (the melody returns following the same patterns, every column in a temple has the same height, etc.). Predictability decreases the amount of information only when a phenomenon is seen as a message. The same rule applies to human cognition: it is the error, the exception to the imagined order of things, that demands concentration and further processing.

Shannon's theory of information, which provides a mathematical foundation for all information technologies, has very important implications for human culture, as it undermines the intuitive assumption that order and predictability are more valuable than disorder and unexpectedness. In the digital realm, it is the other way around: if a sound file is more noisy and nuanced, it is harder to compress for software that automatically finds patterns and redundancies. Reasons for this are conceivable only in probabilistic terms: a pattern is a fixed rule that determines the relation between elements in a system and can be compressed into an algorithm. Grasping this rule allows a cognitive system that perceives a pattern to predict its future development. However, when indeterminacy creeps into the experience and cognitive predictions necessarily fail, sensory inputs have the advantage over perceptual hypotheses. Listeners willing to confront an unpredictable flow of sounds enter this quasi-meditative state of self-alteration; they do not seek recognition and confirmation, but immerse themselves in a flow of new information.

As such, it is no coincidence that both Cage and some contemporary proponents of 'predictive brain' theory in the neurosciences have been drawn to meditation practices (Hartkamp and Thornton 2017; Elk and Aleman 2017). In light of predictive processing theory, zazen meditation works so well therapeutically because it redirects the prediction flow from ultimately isolating self-confirmation to self-alteration. As outlined above, predictive brain theory explains that biological organisms favour accuracy over error, and try to minimise the difference between their models of the future and actual sensory input. To this end, they can either attempt to confirm their hypotheses, or update them. In the first scenario, an organism strives to reinforce its expectations through action. One example of this behaviour would be anticipating the location of a light switch in a dark room, followed by groping around (Friston 2010: 3). The hypothesis is followed by an action that intends to confirm it. Another example would be leaving the concert hall when the performance does not meet one's expectations. In this case, an excess of uncertainty forces the organism to flee. The second strategy consists in transforming the internal model based on acute observation instead of error avoidance. Meditation falls into this category: 'What Buddhism and related meditative practices try to accomplish [is to] focus on simple changes in interoceptive and exteroceptive inputs, without activating higher level conceptual, or even goal-related (cf. "nonjudgmental") predictions about them' (Van de Cruys 2014: 177). A Buddhist monk who is determined to remain motionless for extended periods of time consciously deprives himself of the possibility of confirming his mental hypotheses through action. Instead of following his thoughts, or acting upon them, he is obliged to increase the actual sensory stimuli: 'In this case, the discrepancy between predicted and actual sensory data will be reduced by adjusting the internal model so that the next prediction will be a closer match to the current situation – continuing to sit' (Pagnoni and Hasenkamp 2015).[6] Being unable to escape from sensory signals to the safe haven of the expectation-confirmation dialectic, the meditator is bound to accommodate internal and external information into an updated model of their extended self.

Cage claimed that he 'never practiced sitting cross-legged' (in Jaeger 2013: 50), yet he made references to Buddhist practices frequently. It seems plausible that, in his constant search for novelty and change through art, he was naturally drawn to 'the outward path, rather than the inward path of meditation' (in Kostelanetz [1987] 2003: 52). Meditating indeterminacy did not serve him as a tool for finding peace, but rather for letting go of a fixed identity and embracing endless change (this stance echoes the ideological mantras of the Esalen Institute). Peter Jaeger suggests that 'instead of meditation and the disciplined study of scriptures, Cage used writing, music, visual art, and other forms of cultural production to form a type of meditative engagement with "silence," which he regarded as a "ground, so to speak, in which emptiness could grow"' (2013: 53).

---

[6] Pagnoni and Hasenkamp use the terms 'confirm' and 'revise' to specify two main strategies of maintaining perceptual homeostasis. The first notion describes the activity of pursuing evidence that verifies one's assumptions. The second involves changing one's representations in response to new information.

Cage himself admitted that he used the I Ching as a disciplinary device, and expected the same kind of discipline to 'whatever happens next' from his performers and audiences. In his own words, using 'chance as a discipline' (in Kostelanetz [1987] 2003: 17) allowed him to free people 'from their likes and dislikes, and to discipline themselves' (102). Listeners who could find no discernible relation between the sounds in a tune, or infer the source of a sound during a live performance, were incapable of making accurate predictions. Being unable to infer the distal causes of sounds (for example, note A should follow note B to remain 'in tune'; this melody expresses a feeling of sadness; this sound was caused by air moving through the instrument, etc.), they were forced to constantly update their extended self models to make new predictions.[7]

In Cage's philosophy, the notion of deriving pleasure from the unexpected was thus inseparably connected with the self-discipline of restraining one's egotistic desires. However, I would add that the concept of 'unexpected pleasures', Cage's fetishisation of surprise, stemmed from his strong conviction that information technologies, like the camera, strip humanity of individual intentionality and substitute politics with inhuman forms of governance. What Cage found so liberating about cybernetic technologies, and new media in particular, was their unconditional purposelessness, an opinion he shared with Marshall McLuhan, something that eventually 'infected' and affected the human minds connected to them (McLuhan 1994). Only after taming individual expressivity would it be possible to decentralise space and create a safe environment in which surprise was not associated with danger. Cage's performance as a camera can thus be seen as part of a long list of devices imagined by his friend, Buckminster Fuller, 'that will induce the right behaviors' so that one would not have to 'depend on politically enacted and enforced reforms' (1981: 252). All in all, as Cage courageously demanded in 1966, music, as a political instrument, should 'sing the final dissolution of politics-economics' (1967: 27).

Variations VII: Amplified 'Whatever'

One of Cage's iconic pieces, *Variations VII*, illustrates both the mind-opening and disciplinary effects of Cage's indeterminate performances

---

[7] Of course, they could reject the experience altogether and embark on an inner journey where sensory data perfectly match projections, or by making an infallible hypothesis like 'this is not art', which is confirmed by whatever happens.

on the audience. It was staged in October 1966 at New York's 69th Regiment Armory during '9 Evenings: Theater and Engineering', a festival of experimental art. In the souvenir programme provided to ticket holders, Cage described his idea for the performance as follows:

> My project is simple to describe. It is a piece of music, *Variation VII*, indeterminate in form and detail, making use of the sound system which has been devised collectively for this festival, further making use of modulation means organized by David Tudor, using as sound sources only those sounds which are in the air at the moment of performance, picked up via the communication bands, telephone lines, microphones together with, instead of musical instruments, a variety of household appliances and frequency generators. The technical problems involved in any single project tend to reduce the impact of the original idea, but in being solved they produce a situation different than anyone could have pre-imagined. (Cage n.d.)

Even if this does not sound as simple as Cage promises, one important aspect of *Variations VII* is, in fact, very simple, and bears a relationship to silent pieces like *4'33"* – its (then non-existent) composition contains no clues to indicate what sounds should be played. In a sense, it is a blank page with a mass of endnotes. There is neither a fixed score nor a time limit, only an abundance of technical cues. Most of the devices are not even considered sources for the specific (deliberate) sounds; they are transmission media meant to communicate and alter (unintentional) sounds produced elsewhere. For example, phones used during the performance transmitted ambient sounds from ten places around New York, including the ASPCA dog pound, Merce Cunningham's studio, Luchow's restaurant, the *New York Times* press room and the 14th Street Con Edison electric power station. Magnetic pickups on the telephone receivers conveyed these incoming signals through a complex sound processing system at the Park Avenue Armory. Geiger counters detected ionisation events occurring in the room and produced their characteristic 'clicking' sound. Finally, microphones attached to human bodies also captured internal sounds, like heartbeat and respiration. All this made *Variations VII* a highly complex echo chamber, in which sounds resonated with each other, affecting the bodies of participants who, in return, responded with involuntary noises of their own. For the cherry on this haphazard cake, Cage installed photocells, which automatically turned some of the sound sources on and off. Thus, even if the project was 'simple to describe', in reality its complexity and unpredictability made it impossible for any stable probabilistic model to predict incoming auditory stimuli. A random visitor wandering freely within this mind-boggling technological network was only slightly more clueless than the performers and the composer.

Obviously, there was an explanation attached to the piece that helped to make sense out of all this, but it was almost the same story Cage repeated for most of his pieces. As Douglas Kahn states:

> His idea that all that is necessary for music to exist depends on how one's attention is directed, that is, how one tunes in, is also evident in *Variations VII*. In his notes to the composition, the transduction involved in tuning in becomes a form of fishing, where centripetal musical listening hauls in all sound: 'catching sounds from air as though with nets, not throwing out however the unlistenable ones [. . .] making audible what is otherwise silence therefore no interposition of intention. Just facilitating reception.' (2013: 116)

As in *4'33"*, *Williams Mix* (1952), or almost any piece from the *Variations* series, the only subject of the performance was this way of perceiving sound. If, technically speaking, every composition, play, or even a painting does nothing but 'facilitate reception', then the specificity of Cage's piece lay in its apparent lack of specificity. When the audience entered the performance space, they encountered an opaque system of interconnected devices producing unpredictable sounds. Consequently, any attempt to create an effective prediction model was doomed to failure – the flow of prediction, normally an interplay of correct guesses and new information, could only surrender to a cascade of errors. Creating, confirming and maintaining a representational model of the piece was rendered impossible – there were no intentions or causes to infer, and no apparent patterns to identify. Incoming stimuli did not 'stand for' something else, or, more precisely, listeners were never given the opportunity to 'come up with' a representational model to explain away their cognitive uncertainty. Under such peculiar circumstances, the usual guessing game using the standard cognitive toolkit was defeated: there could be no assuming meaning, intention or tone, all of which can be understood as higher-level modes of prediction. Distorted sounds emerged out of nowhere to form temporary alliances with other sounds and then disappeared without cause. A solid model of representation could not be formed to reduce the number of prediction errors; every sound was shrouded in uncertainty. The impossibility of accommodating incoming stimuli into a model of representation forced participants – those who were willing to conform to Cage's rules – to relinquish any attempts at a 'deeper understanding' and remain on the surface. Only then were they able to 'revise' their predictions at the necessary speed.

This kind of unfocused awareness, which Cage thought was characteristic of the electronic age (in Kostelanetz [1987] 2013: 103), is undoubtedly flat; its value lies in its dynamism and multidimensionality. Because there is no centre of (re)presentation or fixed frame of reference, every experience in the performance space becomes doubly relative: to the spatial coordinates of the listeners, who are free to move around, and to their individual sensibilities. Within this non-representational space nothing seems inevitable, that is, bound by the rules of causality. Newton's laws are suspended and chance becomes tangible. On the one hand, this kind of unfocused, probabilistic awareness of the environment can be seen as the ground zero of human perception, as the 'natural' point of view, divorced from cultural background and individual experience. On the other, it also contains an inhuman quality: the inability to infer distal causes of surrounding events produces an abstract environment and disengages the listener. This is probably why Cage's art often feels cold and emotionless. Confronted with an excess of stimuli that resists easy reduction to predictable (compressed and comprehensible) patterns, listeners are forced to navigate an inhuman landscape of information and prediction. The 'natural' sound Cage liberates from the fetters of egotistic culture is, in fact, alienating and machinic. This makes perfect sense when we recall that his liberation of nature 'in her manner of operation' (Cage 1961: 100) was only achieved by 'dehumanising' tools of indeterminate composition, transmuting familiar, inhabitable space into unfamiliar and inhuman information space. Buried under the layer of meanings tailored for the ego was less the warm womb of Mother Nature than a new world of information technologies. From this perspective, Cage's notion of indeterminate performances can be seen as an early experiment in creating a perceptible information space in which vibrating matter is transformed into probabilistic data, computed by performers and listeners. As the mind extends itself to make suppositions about the world through perception, its presence and action in the performance space transforms it into an information system. Human listeners are reduced – or, depending on one's point of view, elevated – to the role of sensors stimulated by unexpected noise-produced patterns. To begin appreciating indeterminate music, one must identify with the purposeless state of recording devices like cameras and microphones, or, as Cage has put it, one should 'be attentive and empty' (in Kostelanetz [1987] 2003: 239). When one disposes of expectations and the need to confirm them, the image of a meaningful and determinate environment expires. Without stimuli to confirm one's hypotheses, space appears indeterminate, yet remains perceptible.

In a brilliant but overlooked work on digital epistemologies, Tiziana Terranova writes that Western culture owes its informational understanding of space – aesthetically performed by Cage – to electronic media and new discoveries in artificial intelligence. Without these, it was simply impossible to make sense of indeterminate processes:

> Space becomes informational not so much when it is computed by a machine, but when it presents an excess of sensory data, a radical indeterminacy in our knowledge, and a nonlinear temporality involving a multiplicity of mutating variables and different intersecting levels of observation and interaction. Space, that is, does not really need computers to be informational even as computers make us aware of the informational dimension as such. An informational space is inherently immersive, excessive, and dynamic: one cannot simply observe it, but becomes almost unwittingly overpowered by it. It is not so much a three-dimensional, perspectival space where subjects carry out actions and relate to each other, but a field of displacements, mutations and movements that do not support the actions of a subject, but decompose it, recompose it and carry it along. (2004: 37)

To reveal the informational aspect of space, we must remove the representationalist coordinates that govern everyday experience, making the surrounding world more predictable: infer intentions, assign meanings to things and events, make assumptions of purpose, etc. For a person from Western culture, who familiarises these concepts and embeds them in the very fabric of reality, adopting an informational point of view is a violent and alienating experience. I thus find it very confusing that Cage is usually saddled with an interest in the naturalness of sound or the affirmation of life.[8] The 'nature in her manner of operation' reveals itself none too easily – it takes a great deal of self-discipline and reliance on external technologies to experience the indeterminate. Simply put, Cage did not denounce the familiar culture of predictability in favour of an immediate and romanticised nature like many of his contemporaries; he embraced the inhuman future of the emerging techno-culture, which used indeterminacy and unpredictability to produce information. Cage's road to anti-spiritual Enlightenment led through alienation and cycles of 'creative destruction', enriching the informational fabric of posthuman techno-culture:

---

[8] For example, the Wikipedia entry devoted to Cage opens with a statement that suggests this interpretation. https://en.wikipedia.org/wiki/John_Cage. Accessed 22.07.2019.

'Anything resembling an interruption, a distraction, should be welcomed. Why? Because we will realize that by these interruptions and distractions and flexibilities we enrich the brushing of information against information, et cetera' (in Kostelanetz [1987] 2003: 253).

Performance as Randomness in a Jar

The groovy lava lamp, a standard fixture in 1960s America, is a very Cagean set piece. To borrow a term coined by Marcel Duchamp, one of Cage's icons, the lava lamp is 'canned chance' manufactured en masse. An effect called Rayleigh-Taylor instability is responsible for the unpredictable movement of colourful liquids inside the lamp. This phenomenon occurs when a lighter fluid pushes against a heavier one, inducing a turbulent process, so sensitive to minor perturbations that no equipment can predict it. In 1963, Edward Craven Walker came up with the idea of the lava lamp: displaying the erratic dance of gooey fluids by containing them in a heated glass jar, thus 'harnessing' physical indeterminacy for aesthetic purposes (and accidentally providing a source of inspiration for countless acid trips). However, what was merely an amusing gimmick in Walker's day was employed for very serious purposes in later decades by the neoliberal descendants of the California hippies: the 'creatives' of Silicon Valley. Walker, an accountant and a devoted nudist, was probably unaware of the fact that his device – just like Cage's compositions – was also a weird computer. Programmers at Cloudflare, a San Francisco-based cybersecurity firm that provides services to companies like Uber and OKCupid, inspired by Walker's invention, used it as a reliable source of entropy for their encryption software. Their security system consists of eighty lamps placed on standard bookshelves, and a high-resolution camera instead of a stoned teenager as the beholder. The information from the video feed is sent to a server, which encrypts client companies' data. It just so happens that the 'attentive and empty' eye of the camera in San Francisco is never bored by the kitschy performance of the lava lamp's canned chance. Bizarre as it may seem, lava lamp security is virtually unbreakable.

What makes the lamps so reliable as sources of information is not only the endless chemical process of intermingling liquids. The discipline of the glass container, the set of rules governing the relations between agents, and the keen eye for changing patterns are all equally important. Without these extra requirements, liquids would only communicate among themselves; by adding an observer and a set of constraints, the random movement of particles becomes cybersecurity data. The structure (composition) governing the performance of matter imposes an identity on its indeterminate behaviour – neither the changing colour of fluids nor the shape of the lamp determine the exact movements of particles; they only establish a field of possibility. Similarly, the mystical transfiguration of reality through indeterminate performance happens via mediation. My suggestion is that the very concept of performance indeterminate of its composition (or script) is conceivable only if one assumes that matter, when properly framed, is capable of communication, of being transformed into information through predictive action and perception. Moreover, Cage's musical camera-composition not only allows the performance of matter to be framed as aesthetic, it also provides audiences with the stability (discipline) they need to take pleasure in unpredictability. For this reason, it demands absolute submission from the performers. Impersonal and faceless technology that restricts and confines the egotistic drives of the artist creates a safe space, free from individual taste, social bias, cultural preferences, etc. The aesthetic pleasure projected by Cage hinges on the ability to enjoy unpredictable events that, in turn, become possible only in a secure and non-hierarchical environment.

For this reason, any performance theory that draws direct inspiration from Cage's work should take into account both the experimental (systematic) and random (emergent) qualities of his indeterminate performances. The camera depicted in 'Indeterminacy' is thus a universal machine (like the one envisioned by Turing) that translates the disorganised and contingent proximity of objects and actions in space and time into an entangled system. In other words, it turns matter and energy into information. Reality captured with photographic performance is not just *any* material process; however, issues of content and form are completely irrelevant with regard to this definition. Cage's performance comes into being when an activity is captured by an observer (or many observers) with technical equipment (metaphoric or not). For a better understanding, this notion can be juxtaposed with Peggy Phelan's definition of performance in *Unmarked*. For Phelan, performance is precisely what 'cannot be saved, recorded, documented, or otherwise participate in the circulation of representations of representations' (1993: 146). If that is so, then the understanding of performance that emerges from Cage's writings points in exactly the opposite direction. That is to say, even though performance always unravels in the here and now and is never intended as representation (an externalised model of memory), it is constituted as such by the existence of a frame that establishes a space of possibility. In a nutshell, indeterminate performance can be indeterminate only from a perspective that is oriented toward indeterminacy. Experience with a camera in one's hand differs from empty-handed experience:

> The gesture of photographing is one of hunting, where the photographer and the camera unite to become a single, indivisible function. The gesture seeks new situations, never before seen; it seeks what is improbable; it seeks information. The structure of the gesture is quantal: it is one of doubt composed of point-like hesitations and point-like decisions. It is a typically postindustrial gesture: it is post ideological and programmed, and it takes information to be 'real' in itself, and not the meaning of that information. This obtains not only for the photographic gesture, but also for every gesture of every fonctionnaire, be he bank clerk or president. (Flusser 1984: 27–8)

Following Flusser, we can say that in performance the logic of representation (the event stands for something else) is substituted by a probabilistic relation between the actual event and the frame.

Moreover, following both Flusser and Clark, the value of an aesthetic experience's indeterminacy lies not in its unrepeatable quality, but in its relative difficulty of processing and predicting. In the context of Clark's theory, we might argue that Cage's compositions – invented as cameras for capturing contingency – display a certain probabilistic self-awareness. His experimental art – devoid of symbolic meaning, self-expression or other forms of cultural subordination – can be seen as entirely geared against the tendency to reduce cognitive uncertainty and produce the improbable (a different kind of uniqueness than unrepeatability). For a probabilist like Cage, making experimental art was motivated by unconditional affirmation of the here and now and/ or by a rebellious urge to 'rage against order' (Peckham 1965). It was driven by an insatiable curiosity that extended beyond the limits of the human perspective. Therefore, it was no accident that his art was so enthusiastically received at the time by such radical transhumanists as Stewart Brand and John Brockman,[9] perhaps even inspiring their

---

[9] Stewart Brand is a writer, famous for putting together the *Whole Earth Catalog*, the unofficial Bible of the techno-enthusiastic hippies. John Brockman is an author and literary agent, who founded the Edge Foundation, which to this day publishes an online magazine (edge.org) that explores the cultural and philosophical dimensions of science and technology. In his early book *By the Late John Brockman* (1969), inspired by Cage's philosophy, he announced the death of man and argued that all categories that characterise human existence have to be created anew with reference to new achievements in science.

futuristic visions (the latter admitted that he was handed a copy of Wiener's *Cybernetics* by Cage himself).[10]

Cage's novel approach to performance is not only an important example of an intellectual affinity between avant-garde artists and cyberneticians; it also proves that long before both groups started collaborating in New York (see Pickering 2011: 85–9), they shared an anti-representational stance. Moreover, it was the probabilistic epistemology of cybernetics that decisively influenced and shaped the performative turn in the United States. Cage was well aware that the relation between indeterminate performance and technological reproduction was not conflictual. On the contrary, it was unavoidable. It was the very possibility of capturing vibrations of matter and translating them into machinic codes that allowed him to appreciate noise as a 'resource' in art. Cage's path to 'nature in her manner of operation' inevitably led through alienation in technological and probabilistic abstraction. If this technological bias lay at the core of thinking of performance and as performance, uncovering it might prove especially useful in contemporary culture. Does abandoning the codes, values and practices of representation mean embracing an absurd perspective of probabilities? If our performances in the digital age ultimately refer to nothing, should we not only look at their efficiency – 'how they do things' – but also at their (im)probability? As the world designed by Cage's admirers becomes more tangible (Turner 2006), we are slowly discovering the epistemological and political significance of his work. For this reason, I found it important to return to his unexplored, yet prophetic writings, which reveal the transhuman genealogy of the performative turn in the 1960s. Amusingly enough, in our current dark age of social networks and full-blown aestheticisation of politics, Cage's distrust for humanist values (in particular, individuality) and his voluntary submission to indeterminate technological systems make for an aesthetic-political project that deserves serious discussion.

---

[10] Brockman claimed this in an interview included in the movie *Das Netz* (2004).

Chapter 6

Xenakis: Ecstasies of Calculated Randomness

Olivier Messiaen, under whom Iannis Xenakis studied composition in 1951–3, once said of his student's music that it 'is not radically new but radically other' (in Fleuret 1981: 19). It would be difficult to find a higher compliment for an artist as absorbed by the pursuit of making something as unique – or, in his own (unbashful) words, 'different from anything else that has ever been written in the world' (Duffie 1997). Messiaen did not deny the innovativeness of his pupil's ideas; on the contrary, he expressed his absolute admiration for their unprecedented uniqueness in the history of music. The sound of Xenakis's early compositions, created in the mid- and late-1950s, did not resemble anything that had ever been heard in concert halls. In truth, they did not even sound like anything that could be heard outside of them, either. While the aesthetic revolution concurrently initiated by John Cage involved letting some fresh air and street noises into the conservative institutions of Western music through wide open windows, Xenakis's tactics, more subtle, yet comparably alien to listeners at philharmonics, used a traditional means of expression, the orchestra, to produce unprecedented and 'inhuman' sound signatures. Milan Kundera, who called the Greek composer 'a prophet of unfeelingness' provided an evocative description of these soundscapes as 'both objective and unreal, free from the stigma of human subjectivity, aggressive and oppressive' (Kundera 2020: 74). His disturbing soundscapes were a perfect soundtrack for the misanthropy of *homo sovieticus*, for his disappointment with social reality and sense of powerlessness. Kundera reminisces that, during the Soviet occupation in 1968, listening to his recordings gave him a feeling of respite, because 'it talked about the sweet inhuman beauty of the world, before or after human rule'. In the artificial nature Xenakis created, in the buzzing swarms of violins and percussion, you can still hear distant echoes of a familiar world – the rain drumming on the

196

pavement, insects streaming to the hive, or an industrial polyphony of machines operating furiously inside a factory – but even these superficial echoes of musical naturalism usually give way to a completely uncanny mathematical abstraction.

Kundera's reaction to Xenakis's music – the uncanny sensations of 'unfeelingness' or 'strangeness' he claimed to experience – should hardly surprise anyone familiar with his works. Even though Xenakis intended his early pieces to produce an immediate affective response in the listener (Varga 1996: 62) to his waves of ferocious and unsettling sonorities, he approached composition almost like engineering, meticulously calculating the notes. This is not to say that his music was primarily conceptual or experimental. Xenakis felt offended by such suggestions, despite the fact that most of his early texts on music theory read like mathematical textbooks. His music was rooted in concrete, biographical experiences, but, until late in his life, he rarely disclosed how they affected his creative method, because – like many other artists in post-war Europe – he found it impossible to communicate the mental and physical trauma inflicted upon him during the Second World War. When asked if he composed without engaging emotions, he replied:

> Yes, if you mean that kind of traditional sentimental effusion of sadness, gaiety or joy. I don't think that this is really admissible. In my music there is all the agony of my youth, of the resistance, the occasional mysterious, deathly sounds of those cold nights of December '44 in Athens. (In Griffiths 2001)

This is particularly true for the compositions written in the late 1950s and early 1960s, which convey a clear sense of dread and anxiety, and echo the auditory landscape of modern war.

To emulate – or sublimate – these experiences in art, Xenakis turned to statistics and probability theory as tools for modelling stochastic processes, which he wanted to produce as sonic events in concert halls.[1] Xenakis, who graduated from a polytechnic and worked as an engineer, was the first artist to make use of such tools for purely aesthetic purposes, which had been reserved for technicians. This choice was not arbitrary, nor motivated solely by an urge to experiment. Probability calculus allowed him to create mathematical structures for calculating stochastic structures he called 'clouds' or 'masses' of

---

[1] A stochastic, or random, process is a system which undergoes chance fluctuations as it evolves in time, and cannot be fully controlled or predicted, like the stock market or the weather.

sound – sonic entities that he found 'aesthetically interesting' which could have not been produced without this sophisticated methodology (Matossian 1990: 256). On the one hand, they resemble some unspecified natural phenomena – complex processes on the verge of chaos that seem unlikely products of individual intentionality. As such, they can provoke strong emotional responses in the first listeners, overwhelmed by their sheer intensity and complexity. On the other, these unsettling auditory entities never sound completely random, as they are clearly 'defined through global textural and dynamic properties' (Haswell and Hecker 2007: 112). Their identity, secured by mathematical calculation, aesthetically translates into an alien order, obviously structured but hard to identify and follow. Nouritza Matossian, Xenakis's friend and biographer, rightly points out that the uniqueness of his aesthetic project lay in his intent, not as a scientist, or engineer, but as an artist, to master processes of dynamic equilibrium – an idea that had been expressed, for example, by Piet Mondrian in 'Plastic Art and Pure Plastic Art' ([1937] 1964) in the context of painting, or by Edgard Varèse in music ([1933] 2004), but never actually realised.

Xenakis employed probability theory, the best available tool for modelling random processes occurring in nature, precisely to sculpt these forms of dynamic equilibrium. His use of mathematics has been analysed by numerous scholars and the composer himself. Given that most of these analyses focus on the technical aspects of his endeavour, in my reading of his work I want to zoom in on the role and significance of probability in his early works, though in strictly philosophical terms. I will argue that, not unlike the Futurists, Xenakis wanted to bring about a new kind of 'probabilistic sensibility' that finds pleasure in locating a structure that emerges from sonic mass phenomena. Moreover, drawing inspiration from pre-Socratic philosophy and historical analyses of ancient music, primarily Greek, he attempted to give music a different meaning, which I characterise as a sonic representation of *zoe*, a form of undead existence, beyond the distinction between nature and artifice.

Probability and Discontinuity

Although Xenakis spent most of his life in exile, for the most part in France, his work's main reference point remained Greek culture and history: both ancient, as grounds for artistic and philosophical exploration, and contemporary, in the war trauma which marked the composer's life. Born in 1922 in Brăila, a small town in eastern Romania,

Xenakis came from a relatively wealthy mercantile family. After his mother's premature death, when Iannis was only five years old, his family returned to their homeland, where he remained until his flight to Western Europe in 1947 due to the political situation in Greece. After the retreat of Nazi troops, deep social divisions and conflicts rooted in the pre-war struggle between conservative monarchists and progressive republicans resurfaced, tearing the fragile society apart. The arrival of the British army in Athens in October 1944 did not bring long awaited stability; many Greeks who hoped for a democratic revolution viewed it as the substitution of one occupier by another. Xenakis, who initially enlisted with the National Liberation Front, a left-wing resistance group, took active part in guerrilla warfare against the British troops in Athens. During street battles he was hit by shrapnel from an artillery shell, losing one eye, leaving the left side of his face permanently paralysed, and barely making it out alive. After recovery, Xenakis returned to the university to obtain his diploma in engineering, and then – to avoid being locked up in a concentration camp or even facing the death penalty – he escaped to the West, first to Italy and then to France (for a detailed account of Xenakis's early life see Matossian 1990: 21–43).

Undoubtedly, his quasi-scientific project of introducing probability theory to musical composition was not necessarily causally linked to his traumatic experiences, though it is important to take note of this important biographical factor. For Xenakis, science was less an instrument than a spiritual activity. Deeply affected by war and with no clear sense of belonging to any particular cultural tradition, he clung to a belief that only through science would it be eventually possible to 'achieve universality':

> For years I was tormented by guilt at having left the country for which I'd fought. I left my friends – some were in prison, others were dead, some had managed to escape. And I felt I had a mission. I had to do something important to regain the right to live. It wasn't just a question of music – it was something more significant. (. . .) I became convinced – and I remain so even today – that one can achieve universality, not through religion, not through emotions or tradition, but through the sciences. Through a scientific way of thinking. But even with that, one can get nowhere without general ideas, points of departure. Scientific thought is only a means through which to realize my ideas, which are not of scientific origin. (Varga 1996: 47)

These three topics – war trauma, science as a universal language, and art's higher purposes – often intersect in conversations with Xenakis, as well as in his technical and philosophical essays. Science, and mathematics in particular, often serves as a defence mechanism to sublimate his traumatic experiences – from a near-death accident to a betrayal by his own fellow citizens – and to avoid directly engaging with his personal emotions. Hence, Xenakis's view on science's universalism differs sharply from classical and positivist doctrines that imagined scientific methods would replace the metaphysical constructs that form the foundation of human knowledge and social practice. Universality imagined by Xenakis was not a positive or definite discourse of universally true statements to dictate the rules for establishing personal or collective utopias; it was ambiguous and alienating. Not only was it devoid of certainty, it was also removed from an individual human perspective. Xenakis placed new probabilistic sciences at its heart: an unusual idea at the time because probabilistic theories, like the substitution of easily fathomable atomic orbitals for probability transitions in quantum mechanics, were generally considered to be either imperfect approximations, and thus weak and suspect foundations for any system of knowledge, or as operative but confusing models completely alien to a human experience of reality.

This widespread sentiment was nicely captured by Simone Weil, who had a good grasp of the modern sciences, and was primarily interested in its epistemological complexities and how they translate into existential considerations. In a late essay titled 'At the Price of an Infinite Error: The Scientific Image, Ancient and Modern' ([1941] 2015), she compared three Western epistemologies of scientific thought in the light of their anthropological implications: ancient (Greek), modern (Newtonian) and contemporary (Copenhagian). While she had a great deal of praise for the first two paradigms, her attitude toward the third was purely critical, as she laid out in the first sentence: 'Something happened to the people of the Western world at the beginning of the century, something quite strange: we lost science without even being aware of it, or at least, what had been called science for the last four centuries' (156). In her view, the scientific community passed the point of no return in 1900, when Max Planck introduced his theory of quanta and probability was introduced as a key concept in natural sciences.

Weil's criticism pertains solely to the existential aspects of science: where it situates the human being in relation to the universe it tries to explain. Most of her argument revolves around crucial differences between world-images enclosed within classical and quantum mechanics. The former, she argues, takes the notion of work (and, consequently, energy) as a key concept in its pursuit of a fixed and stable representation of the universe. In doing so, science speaks of phenomena that have important existential significance for people: it studies 'geometric and mechanical necessities' which dictate the conditions in humankind's struggle for survival – 'the curse of work' brought upon it after the expulsion of Adam and Eve from Paradise (158). Classical science guides in practical matters and uncovers a world that is ultimately cold and indifferent to human beings. Yet, even if it is neither good nor bad, neither beautiful nor ugly, it does give us firm ground to stand on.

This clarity collapses with the advent of quantum physics. Weil assigns a special role to Planck's theoretical intervention, which breaks down electromagnetic action, or the loss of energy during radiation, into 'portions', or minimal amounts of energy, thus introducing an essential discontinuity in our understanding of physical reality. Weil argues that, from the standpoint of classical physics, this hypothesis was unthinkable '[f]or energy is a function of space, and space is continuous; it is continuity itself; it is the world thought from the point of view of continuity; it is things in general insofar as their juxtaposition envelops the continuous' (174). Weil is also concerned by the fact that Planck's constant and formula were products of pure theoretical speculation: its 'provenance one cannot imagine and a number corresponding to a probability, has no relation with any thought' (174). Planck himself shared Weil's reservations and was puzzled by the fact that his mathematical model produced accurate predictions in experimental contexts. Almost thirty years later, in a letter to an American physicist named Robert Wood, he confessed that 'this was a purely formal assumption' (Kragh 1999: 62) and that, at the time, he had no idea this would necessitate a break with classical physics. Moreover, he conceded he had used statistical mechanics 'as an act of desperation' because he did not agree with Boltzmann's statistical interpretation of thermodynamics. Like Weil, Planck revolted against his own idea, afraid of its further implications and appalled by its aesthetic form.

Weil, who had studied Planck's writings in detail, took note of his unconventional method and presented it in terms of a scientific scandal. In her eyes, not only did theory of quanta destroy the concept of continuous space, but, as a result, it placed probability at the heart of modern scientific discourse, obfuscating explanations about the universe that used to be clear, precise, and definitive:

> Discontinuity, number, smallness, that is enough to give rise to the atom, and the atom has reappeared in our midst with its inseparable train of retainers, that is to say, chance and probability. The appearance of chance in science has been a scandal; we want to know where it came

from, yet only have to reflect that the atom brought it; one only has to remember that already in the ancient world chance went along with atoms, and one has never dared to think that it could be otherwise. (175)

Weil was concerned by the inverted logic that engendered the weird, discontinuous image of reality in quantum theory. Quanta were not real entities observed in a laboratory and then explained using theoretical apparatus – they emerged out of theoretical speculation only because a scientist resorted to probability theory for lack of a better solution. Regardless of Weil's assessment, she rightly noticed that the introduction of probability theory in physics also entailed the concept of physical discontinuity and led to the erosion of meaning (referentiality) in science. When science began formulating its statements as mere predictions, its symbols no longer indicated tangible entities that existed in reality, because the distinction between the subjective experience and objective conditions collapsed. Interestingly, as Weil observes, not only objectivity, but continuity of physical space was lost in the process.

During his long career as a composer and sound engineer, spanning from the 1950s to the end of the twentieth century, Xenakis took up and reflected upon all of the topics Weil mentioned in her essay. Like the French philosopher, he was convinced that the arrival of probability theory in modern science was neither accidental, nor simply a methodological issue, relevant only to a small group of practitioners in theoretical physics faculties. And like Weil, he was conscious of the fact that this new image of physical reality was not only essentially predictive, but also grainy and fragmented. In short, Xenakis was well aware of the connection between probability and physical discontinuity; but unlike Weil, he did not reject this conclusion; he wanted to give this new image an aesthetic and affirmative form. It was no accident that the application of probability theory in composition eventually led him to research quantified sound, called granular synthesis, which he undertook on a few occasions and in several research groups from the 1960s onwards (mostly at Centre d'Etudes de Mathématique et Automatique Musicales in Paris). This theory, alongside his invention of UPIC Workstation (Unité Polyagogique Informatique CEMAMu), a music composition tool in the form of a digitising computer graphics interface,[2] might have turned out to

---

[2] This device made it possible to draw arc-shaped forms on a simple electronic touchscreen, representing the pitch and duration of computer-generated sonic events. They could be assigned to any soundwave samples and layered on top of one another, making it possible to design 'clouds' of sound, as much as sixty-four grains in density.

have the greatest impact on modern music. Even though the machine never achieved commercial success, it inspired further research into the synthesis of microsound using statistical and probabilistic methods, which produced analogue devices and numerous digital programs still used in electronic music (from ambient to techno).

In his aesthetic and quasi-scientific investigations, Xenakis drew inspiration from a formal system proposed in 1946 by Dennis Gabor, a Hungarian physicist most renowned for his research in optics, particularly holography, for which he was awarded a Nobel Prize in 1971. Gabor's early speculations on the nature of sound, inspired directly by quantum physics, were never experimentally proven and were subsequently overshadowed by his other scientific achievements; nonetheless, they inspired Xenakis to work on his own system of synthesising elementary sounds from even smaller units. Gabor proposed to use Fourier analysis, which breaks down continuous functions used to represent sound waves into a sum of sine and cosine waves (1946). Such a method had been used before him, but only for waves of constant frequency, rarely to be heard in real life. To sidestep this problem and apply the same tools to study natural sounds, Gabor proposed that every complex sound, just like any other transformation of energy, can be approached in terms of quantum mechanics and divided into smaller units: quanta. In making this assumption, he did not want to prove the existence of a new physical molecule, but to hint at the possibility of theoretically fragmenting sound into hypothetical microwaves, whose properties could be determined statistically: by the Gauss function of time and frequency. In some respects, Gabor's concept resembles the theory of phonons – hypothetical quasiparticles, quanta of energy in vibrating boson networks. The similarity between these ways of understanding sound, however, does not go beyond a belief that it is possible to distinguish and quantify microscopic acoustic events. Gabor did not go so far as to formulate any metaphysical claims. His theory only offered an alternate formal system for recording and producing sounds. He was not interested in acoustics, that is, in physical processes that could supposedly take place in the absence of an observer, but in communicated (perceived) sound, i.e. sonic events decoded by a hypothetical listener or a technical apparatus. Therefore, the lion's share of his arguments was mathematical, aimed at supplementing the Fourier model, and theoretical reflections on the systematisation of communication. Gabor's sound unit was, in fact, a unit of knowing. It is Gabor's formalisation – and indirectly, the atomisation – of sound that opened up the opportunity to think about music and create it using statistical and probabilistic concepts and formulae. Earlier on, even before the information revolution of the 1940s, classical composers were sentenced to rely on traditional notation, which not only offers little room to manoeuvre in terms of tonality, but also forces the composer to set predictable sounds (music) against chaotic noise. The quantification of microsounds allows, in turn, for music to be made with noise, that is, for acoustic phenomena to be intentionally organised, effectively abolishing the music-noise duality. This explains why Xenakis's pieces turned out more popular in concert halls and classical music festivals than Cage's or Luigi Russolo's revolutionary ideas: only he had managed to 'rein in chance' and deliver a very complex but still predictable and reproducible sound. It is no wonder, then, that his writings scarcely reveal any appreciation of randomness for its own sake.

Likewise, Xenakis never suggested he wanted to investigate sound in its truest nature. His interests lay in the epistemological aspect of sound perception and, in many respects, resonated well with the philosophy of science formulated by the Copenhagen School. The metaphysical dimension of his aesthetic research and theory consists in an assumption of the fundamental 'discontinuity' of sound and its divisibility into small units of energy, which is possible only when the energy itself 'divides' into particles with certain properties. The capacity of the hypothetical listener of Xenakis's music to distinguish blurred contours of order emerging from chaos depends on the ability to sense those microscopic events that become dissembled and reassembled in the brain. Put simply, the cloud does not reach the listener's ears as a ready-made sound object, it is assembled from sensory data in those areas of the brain responsible for analysing auditory information. We do not identify a crash of thunder as such, nor any other discernible sound associated with an event; it emerges in cognition, out of a sea of data, examined in search of patterns. Xenakis is not interested in sound-in-itself but in sound communicated – resonating in the universe and perceived by an intelligent being on Earth.

From Xenakis's perspective, the greatest merit of Gabor's theory was that it suggested treating even the simplest acoustic phenomena as mass events, allowing single notes to be regarded as entities emerging out of chaos of even tinier swirling particles (or grains). Using these grains, it was possible to create new sounds that did not fit any existing musical notation system. In the 1960s, Xenakis used this method to compose *Analogique A-B,* and in the following decades he got involved in various research projects in which electronic synthesisers were used to produce theretofore unknown sounds. These investigations – not conducted individually, but with teams of physicists and engineers – resulted directly from the vision of music contained in his first mature piece, *Metastaseis.*

Calculating Sounds

Metastaseis was the first work in which Xenakis attempted to mate-rialise his ambition of creating clouds of sound, and this work is widely considered his first major composition. Initially, Xenakis planned for it to be included in a triptych for choir and orchestra called *Anastenaria*, named after a Dionysian ritual that is still taking place in Greece and Bulgaria (on the pre-modern sources of Xenakis's music see: Turner 2005). I will return to the significance of this fact later. The title of the piece is a neologism which loosely translates as 'beyond stillness' (*meta* – after, beyond, *stasis* – immobility; or simply 'transformations' [Matossian 68]) and hints at the composer's pecu-liar goal to organise sound unlike the Western traditions of tonal and serial music. Despite the essential difference between tonality and serialism, both methods of composition are ultimately based on a linear conception of time, dictating the progression of pitches. In Xenakis's view, these predominant Western methods of composition expressed a cultural belief in the uniform and even flow of time in one direction. To escape this narrow concept of temporality, Xenakis decided to focus on the texture of sound – its material presence in space – in place of the overall musical structure. As Matossian reports, instead of focusing on 'traditional formal elements of fig-ure, theme and melody', he was drawn to concepts like '[c]hanges in number or mass, in the play of densities and distribution', which allowed him to 'articulate the variations in fields of sound' (71) and compose an organised cacophony of sounds in which the particular succession of notes was less important than the overall sonic effect.[3]

The piece begins with a single note, played quietly for over a min-ute by the whole string section. As the sound slowly gets louder and louder, individual violins slowly recede into the background through

---

[3] This was a lifelong dream of Edgard Varèse, another French composer, who in 1937, speculated about the future possibilities of composing 'zones of intensity' of 'various timbres or colors and different loudnesses' (. . .) However, Varèse believed that this treatment of sound would only become possible using the electronic equip-ment of the future. 'When new instruments will allow me to write music as I con-ceive it, the movement of sound-masses, of shifting planes, will be clearly perceived in my work, taking the place of the linear counterpoint. When these sound-masses collide, the phenomena of penetration or repulsion will seem to occur. Certain transmutations taking place on certain planes will seem to be projected onto other planes, moving at different speeds and at different angles. There will no longer be the old conception of melody or interplay of melodies. The entire work will be a melodic totality. The entire work will flow as a river flows' (. . .).

slow glissandi – a musical technique of gliding between pitches and playing all the sounds in-between them. The auditory space is soon filled with an immense cluster of sounds across the full range of the orchestra, occasionally interrupted by percussion. James Harley reported that '[t]he impact of this opening passage cannot be over-stated; nothing like it had ever been heard before. For the audience at its 1955 premiere in Donaueschingen, it was as if they were hearing "atomic music" from "the first traveller in space"' (9). After such a dramatic opening, the following sections maintain the intensity and unusual sound signature, but achieve similar effects through differ-ent means. In the next part, for example, Xenakis resorts to more traditional methods of composition, but he splits the composition into different subdivisions of the beat, and '[t]he result is that, in spite of the strict organization of the music, there is a certain "statistical" quality that nudges the contrapuntal nature of the music toward a more textural character' (10).

However, the first piece in which Xenakis actually made use of sta-tistics was his next big composition, *Pithoprakta*, which is another neologism, this time meaning, 'actions through probability'. Composed for the same orchestra as *Metastaseis*, the piece superficially resembles the one before it, yet in terms of its creative method it was an important breakthrough for the composer. It is here that Xenakis finally succeeded in using statistics and probability calculus to 'control continuous trans-formations of large sets of granular and/or continuous sounds' (16), effectively composing whole sections of the piece with mathematical formulas in hand. In *Formalized Music*, he described his method in great detail and specified that statistical mechanics had provided him with both the mathematical tools (Gaussian statistical distribution) and the theoretical concepts (Brownian motion) that inspired him. Concepts and formulas, used on a regular basis to describe consistencies observed in various data sets and predict the distribution of features in large populations, were used by Xenakis to generate values of musical data based on arbitrarily chosen initial conditions.

In preparing the piece, he treated every instrument in the orches-tra like a gas molecule in a cloud where individual particles move within certain limits (dictated by the gas temperature) that determines cloud's identity. Molecule-instruments in the cloud behave according to the Maxwell-Boltzmann distribution law, which explains tempera-ture, a thermodynamic property of a physical system, as a measurable macroscopic result of the microscopic movements of molecules. In other words, from a statistical mechanics perspective, a gas tempera-ture's value, though measured and perceived by a distant observer as relatively stable, results from a chaotic microscopic process. To create his clouds of sound, Xenakis devised a complex method for calculating the state of molecules – their pitch – at any given moment, hoping to create a state of dynamic equilibrium in music. To this end, he adopted and translated concepts borrowed from thermodynamics into musical terms, thus introducing brand new terminology to the art of composition.[4] In the piece's description, he referred to two concepts absent from classical notation: 'speed' (not to be confused with tempo), meaning the distance between pitches, and 'temperature' understood as the (periodic) state of equilibrium between density, dynamics and timbre (for a detailed analysis, see Arsenault 2000). These partial velocities were to make up a specific temperature distributed to follow the Gaussian function (represented by the characteristic 'bell curve').

Every section of the piece was composed using different methods; probability calculus was applied most extensively in a short fragment, only seven measures long (between 52. and 59.). The number of steps and restrictions Xenakis set before himself leaves no doubt as to the discipline of his creative method. After a Sisyphean labour of reconstruction, based on the official score, notes left in theoretical writings and interviews, Linda Arsenault managed to prepare the following algorithm:

> It is possible to compose stochastic music in the style Xenakis created in measures 52 to 59 of *Pithoprakta* provided that: a sense of rhythm is eliminated; the musical elements of duration; density, dynamics and timbre are constant; the pitches are determined randomly, based on numbers generated from a Gaussian distribution with a mean equal to zero and a standard deviation of approximately 25 semitones; no two same pitches occupy the same place at the same time; the pitches are realized by a variety of groups of stringed instruments playing pizzicato glissandi; the pitches stay within a range of 30 semitones for each group of instruments. (57)

To determine the pitches, Xenakis, like Cage, consulted a mathematical oracle: a random number generator. He decided to partially cede control of the score to instruments of chance, because the succession of individual sounds in *Pithoprakta* was irrelevant to the overall musical effect. Instead of precisely determining individual sequences of notes, he invented a system of conditions by which the overall

---

[4] Although these concepts did not gain much popularity among other composers, Xenakis's theory of granular synthesis (1959) received a great deal of attention from sound engineers, programmers and electronic musicians.

sound structure could emerge: higher-order units of sound were to arise from sound waves emitted by musicians whose individual instructions displayed no intelligible order.

Why did Xenakis go through all the trouble of endless calculations to generate a few lines of notes? One reason is that he saw writing music as first and foremost an intellectual and spiritual activity, 'a mission . . . something important to regain the right to live' (in Harley 230) – an anachronistic and exalted attitude among fellow engineers, on the one end of his social spectrum, and avant-gardists and experimentalists on the other. In his philosophical treatise 'Towards a Metamusic', published in 1971, he condemned one group among the latter, whom he called 'today's technocrats'; despite using similar concepts and mathematical formulas, they made the mistake of striving for 'objective and scientific criteria of aesthetic value' (180). Although he did not specify who these technocrats were, one might surmise that he had in mind the artists and philosophers associated with the computer art movement whose interests in electronic technologies differed significantly. The aesthetics of Max Bense, a German physicist-turned-philosopher at the University of Stuttgart, is one particularly great example of such tendencies in post-war Europe, exerting a profound influence on numerous conceptual artists throughout the continent, especially in Germany and Yugoslavia. In 1958, he opened a small exhibition space within the university confines, *Studiengalerie des Studium Generale*, where, until its closure in 1978, ninety-one exhibitions took place, most of them thematically related to new information technologies (Klütsch 2012). Bense, and to some extent the whole of the 'Stuttgart School', was very well educated in mathematics, which he used in solving problems in art. He was interested in finding firm foundations and precise methods for the abstract analysis of art – from semiotic, metrical, statistical or topological perspectives – and for the systematic production of new works. The cornerstone of his grand project was a quasi-probabilistic formula borrowed from David Birkhoff, defining aesthetic value in terms of its relative unpredictability, as a ratio between order and complexity (1933).[5] His goal was to develop a theory that would allow him to measure the amount and quality of information in aesthetic objects, yielding an evaluation of art that went beyond the discourse of traditional art criticism.

---

[5] In line with Bense's theory, Frieder Nake, his student and one of the most prominent artist members of the Stuttgart School, put forward a theory of 'probabilistic decision making' as 'a computable counterpart of intuition' (Klütsch 2012: 75). Nake's dissertation in mathematics investigated a facet of probability theory.

In contrast to Bense and the information art movement, Xenakis paid a great deal of attention to the cultural ramifications of his creative method – its place in the history of music and its broader existential implications. For example, in the opening of his theoretical manifesto 'Free Stochastic Music', he makes the following statement:

> Art, and above all, music has a fundamental function, which is to catalyze the sublimation that it can bring about through all means of expression. It must aim through fixations which are landmarks to draw towards a total exaltation in which the individual mingles, losing his consciousness in a truth immediate, rare, enormous, and perfect. If a work of art succeeds in this undertaking even for a single moment, it attains its goal. This tremendous truth is not made of objects, emotions, or sensations; it is beyond these, as Beethoven's *Seventh Symphony* is beyond music. This is why art can lead to realms that religion still occupies for some people. (1992: 1)

Taken out of context, this reads like an excerpt from an essay by a Romantic artist raving about the mystical power of great art, irreducible to its material form or sensory experience. Although it seems like a textbook example of Romantic idealism, the composer clarified that music, despite its mystical power, developed parallel to other intellectual attempts 'to explain the world by reason'. The apparent contradiction between these two approaches evaporates when we consider that Xenakis understood the experience of 'total exaltation' as suppression of individuality and not reason. On the one hand, he considered music-making to be a rational activity, an operation creating rules that translate into sonic forms. On the other, he concurred with the Pythagoreans, who 'used music to cleanse the soul as they used medicine to cleanse the body' (202). He suggested that music's mystical power to intoxicate the listener, or to 'transport you from one state to another . . . [l]ike alcohol . . . [t]he power of Dionysus', stems from its abstract nature, allowing one to transcend the everyday perspective of one's ego'. Even if a piece of music speaks of emotional experiences, it has to be translated into a system of rules (executed by the orchestra) that carry no meaning at all. Composers (as envisioned by Xenakis) and scientists alike are preoccupied with producing forms that are devoid of all meaning tied to an individual experience of reality. Writing music and the scientific study of physical phenomena share one essential quality: both result in transcending a personal point of view.

In 'Towards a Philosophy of Music', Xenakis refers to the Orphic doctrine which 'taught that the human soul is a fallen god, that only *ek-stasis*, the departure from self, can reveal its true nature' (1992:

201). He does so to distance himself from a reductionist understanding of reason, and reasoning that mistakes thinking for logic – a mere operation on sequences of symbols. Xenakis's enthusiasm for the sciences was countered by his distaste for the forms of linear thinking he found present in all the prevailing methods of composition in the European tradition, such as tonality, serialism or even polyphony. For example, his critique of polyphony, expressed as early as 1955 in the provocative essay 'The Crisis of Serial Music', addressed the problem of its deterministic logic and composers' focus on the individual progression of notes. In polyphonic compositions, based on the simultaneous development of individual melodies, there is no room for true, that is, irreducible, emergent complexity, as the listener can easily break down the overall texture to its constituent elements. For Xenakis, European music had been dominated by this paradigm, which projected individual experience of time onto musical structures, its form thus objectifying a mode of experience focused on the individual life. In his view, art should help to transcend this limited perspective instead of reinforcing it.

It is thus important to stress that Xenakis explained his preoccupation with stochastic processes and probability theory in music with reference to post-Newtonian science, as well as to certain mystical and ritualistic traditions. As I have mentioned, Xenakis paid tribute to Dionysus in his first composition, titled after Anastenaria, a festival that still takes place in a few villages in northern Greece and southern Bulgaria, initially in honour of the Greek god of wine, before being appropriated by the Christians. The choice of Dionysus as a patron is significant, as he was not the only god of music in the Greek pantheon. The other two, Apollo and Hermes, on the surface would have certainly been a better fit for an engineer and serious composer writing music for middle-class audiences sitting politely in their seats in the philharmonic hall. Hermes was also a god of merchants and – as the messenger of the gods – of communication as such (see Serres 1982), whereas Apollo is commonly associated with order and reason. Of the three, Dionysus seems the least probable patron for an engineer and composer who approaches music in a manner not unlike how he calculates the volume of concrete necessary to support a building.

Dionysus, for some reason not mentioned by name, appears in 'Towards a Philosophy of Music': he is the fallen god, worshipped by the Orphists, who gave life to humans. According to the myth from which Xenakis drew, Dionysus, the son of Zeus, was torn to pieces and devoured by the Titans. To avenge his son, Zeus killed the Titans and turned their bodies into the ash from which humanity was eventually born. Apart from serving as an origin story that ties mankind to the gods, the myth also expresses the Greek view of the nature of life as indestructible. The mythology surrounding Dionysus reveals the complexity and intricacy of the Greek understanding of nature, not only among groups of philosophers, but in society as a whole. Contrary to popular representations of Dionysus as the cheerful god of wine, he was often depicted as a violent and unpredictable deity, ready to take the life of anyone who dared to be sceptical of his mad truths (like Pentheus in Euripides' *The Bacchae*, torn to pieces by the god's servants). According to Karl Kerényi, Dionysus was celebrated as the god who symbolised the experience of life in its totality, *zoe* – life as such, existing beyond differentiation into species or individual organisms. Contrary to *bios*, the singular life, *zoe* could not be killed, only transformed into another form, re-entering the Circle of Life. In this sense, the notion of *zoe* expresses an idea similar to the thermodynamic law of energy conservation, which states that energy cannot be destroyed, only transformed. But *zoe* was not merely a theoretical concept – it was also a perspective for formulating one's existence in relation to other living beings. By freeing oneself from the solitary confinement of *bios*, one could re-enter the flow of ever-transforming matter and perceive it in its irreducible ambiguity, untainted by subjective demands and expectations. One way of experiencing *zoe* as a mode of living was through rituals celebrating Dionysus's violent death and rebirth. The Dionysian myth – as reconstructed by Walter F. Otto – celebrated this entanglement of life with death:

> The more alive this life becomes, the nearer death draws, until the supreme moment – the enchanted moment when something new is created – when death and life meet in an embrace of mad ecstasy. The rapture and terror of life are so profound because they are intoxicated with death. As often as life engenders itself anew, the wall which separates it from death is momentarily destroyed. (Otto 1965: 137)

The only way to truly experience *zoe* was through abandoning *bios*, which was preoccupied with the preservation of the individual organism, and sharing the god's madness. This affirmative yet critical dimension of the festival captured the attention of Friedrich Nietzsche, whose *The Birth of the Tragedy* praised the Dionysian religion as a fascinating example of a cultural practice that did justice to life's joyful and violent complexity. From the perspective of Dionysus, or, as Nietzsche calls it, in a state of 'Dionysian arousal' (Nietzsche [1872] 2000: 50), it becomes possible to experience the world beyond 'Oedipal' individuation, as the boundaries of subjectivity dissolve, and the very distinction between subject and object

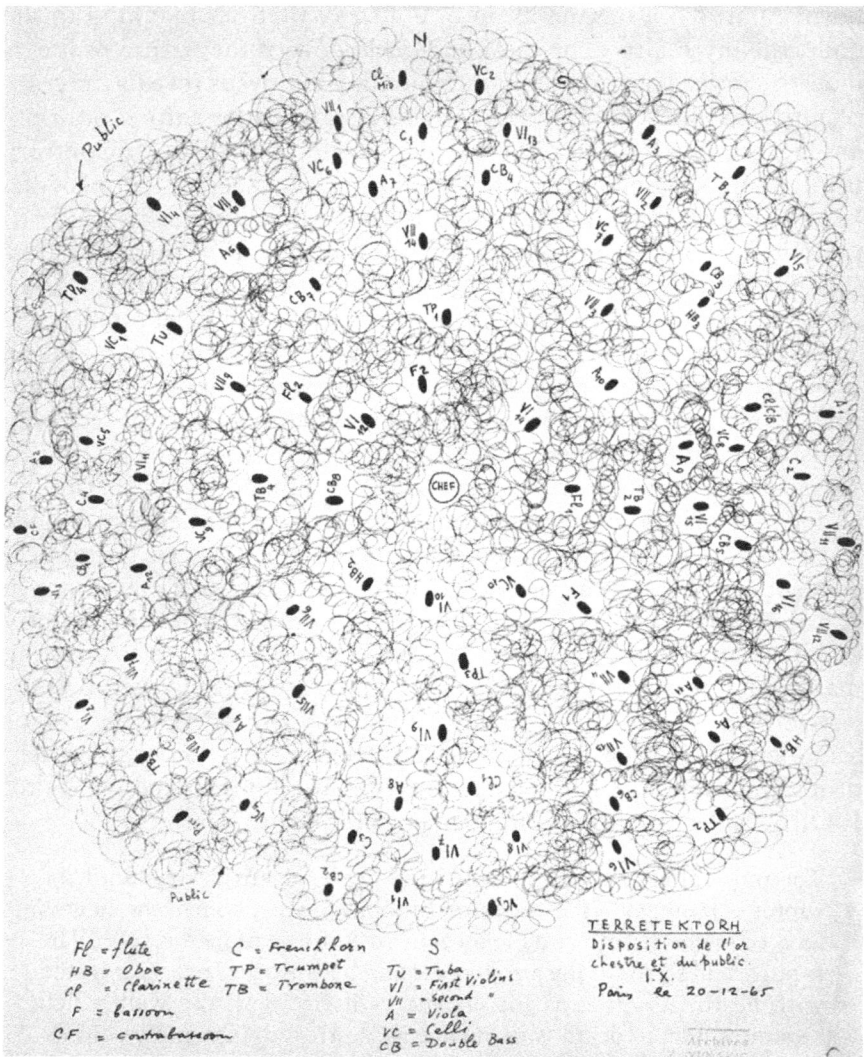

Figure 6.1 Nebulous forms representing the distribution of musicians in concert space. Iannis Xenakis, Study for *Terretektorh*. 20 December 1965. Ink on vellum, 9 × 11 inches. © Copyright Iannis Xenakis Archives, Bibliothèque nationale de France, Paris

falls apart (Cox 2018: 26). In this state of self-annihilation, one is transformed and entangled with the social sphere like bees in a hive (Dionysus was often represented with swarms of bees) – as part of a social mass, a truly statistical phenomenon. Nietzsche even uses a microbiological analogy: upon entrance 'into an unfamiliar nature', a person becomes one with a phenomenon 'in epidemic proportions;

a whole crown felt itself enchanted in this way' (50). The great festival of Dionysia was not a cheerful celebration of nature's fertility, but a perplexing feast in which joyful ecstasy mixed with panicked terror, life's persistence with destruction, absolute closeness with nature with infinite alienation (Otto 92–4) – a mix of notions and affects that properly fits the description of Xenakis's stochastic aesthetics. It is worth noting that the Dionysian assertion that life is not contrary to death was an important cornerstone of his world view: 'Our organism, degenerating every second, knows it. Now, this definitive disappearance can be transposed in the domain of work: the choices that I make when I compose music, for example. They are distressing, for they imply renouncing something. Creation thus passes through torture' (Xenakis 1987: 45).

There is another interesting aspect of the Dionysian philosophy of life: the definition of *zoe* as life at its bare minimum, perceived from an indifferent distance, brings to mind how modern science conceptualises life in terms of chemicals and molecules (Fenyvesi 2014: 52). Kerényi writes: 'For the present-day student of the phenomenon "life," the fact that *zoe* is experienced without limitations is only one of its aspects, not the whole. [. . .] *Zoe* is the minimum of life with which biology first begins' (Kerényi 1976: xxxiv). The ancient Orphists who revered Dionysus and Persephone above the other Olympian gods and modern scientists may not share many obvious similarities – Dionysian madness and scientific rigour seem like opposing standpoints – but both approaches conceptualise life as a physical, systemic process which cannot be grasped by focusing on individual organisms. This is, of course, a far-fetched comparison, which purposefully ignores the striking difference between ancient and modern discourses on life. However, it is no accident that Xenakis cited Orphism and Pythagoreanism – a philosophical school deeply influenced by the cult of Dionysus – as the traditions closest to his broad understanding of 'reasoning', which saw powerful, mystical experiences as legitimate cognitive methods. By referring to these ancient concepts, Xenakis validated his use of science in art and his unique interest in applying probability and statistics to create quasi-ritualistic music. For some it might be a fascinating paradox, for others, a glaring contradiction that Xenakis employed such highly sophisticated mathematical machinery to examine the basic, even primordial capacities of the human brain. On the one hand, he dreamt of the pre-individual, primordial experience of ecstasy; on the other, he remained firmly entrenched in the modern, European paradigm of the grand 'auteur' who tamed chance with probability and exhibited it on beautiful concert hall stages all around the world. Xenakis, to my knowledge, never explicitly addressed this concern, but it should be noted that the stochastic period in his music came at a time when counterculture movements were gathering momentum, and even though he never contested official institutions, his primordial aesthetics could be considered a manifestation of a broader tendency to go back to the roots of culture. If this is true, stochastic aesthetics were always at odds with the official spaces of philharmonics. This is probably why Xenakis was so keen to engage in projects held outside of the confined spaces of highbrow musical institutions.

Probability versus Individuality

Xenakis once said that his listener 'must be gripped and – whether he likes it or not – drawn into the flight path of the sounds, without special training being necessary. The sensual shock must be just as forceful as when one hears a clap of thunder or looks into a bottomless abyss' (in Ross 2007: 300). He also stated on many occasions that stochastic music was born out of conviction that 'clouds' of sound can induce a strong affective response in the listener, who should be able to appreciate new, yet somehow familiar forms of order. Benoit Mandelbrot, a mathematician who dealt with sciences of chaos throughout his career, used to say that 'clouds are not spheres [and] mountains are not cones', implying that the Western intellectual tradition was oddly focused on simple forms rarely found in nature. If human beings are capable of observing clouds moving in the sky and predicting their behaviour in the near future, or determining whether a swarm of bees is approaching or moving away, then it is also possible, from a cognitive standpoint, to discern patterns in semi-random clusters of notes. If one enjoys listening to the raindrops hitting a windshield or finds something sublime in the murmur of a crowd in the street, one also possesses the cognitive resources required to process similar data in a concert hall.[6] No special training or musical education is necessary

---

[6] The same goes for the use of mathematics. The basic mathematical concept used to create *Achorripsis* was the Poisson distribution, which is used to determine the probability of events happening in a sequence independent of each other if we know their frequency. This is generally used by scientists and engineers to model mass phenomena whose elementary events have a constant probability of occurrence (in space and time), for example, the number of mutations in a DNA fragment caused by radiation exposure, or the number of soldiers in an army kicked to death by a horse (assuming that the numbers of cavalry troops is constant). Xenakis used the Poisson equation not to analyse existing data, as is usually the case, but to create a musical structure that would display characteristics of the Poisson distribution.

to feel the immediate impact of sound as a material force. Stochastic music, at least in theory, is essentially democratic. Xenakis hoped his ideal listener would take up the challenge posed by his obscure sonorities and quickly develop a kind of probabilistic perception and appreciation of stochastic aesthetics:

> In fact, the data will appear aleatory only at the first hearing. Then, during successive rehearings the relations between the events of the sample ordained by 'chance' will form a network, which will take on a definite meaning in the mind of the listener, and will initiate a special 'logic,' a new cohesion capable of satisfying his intellect as well as his aesthetic sense; that is, if the artist has a certain flair. (Xenakis 1992: 37)

If so, then such probabilistic compositions speak directly to the human subconscious and the intuitive capabilities which allow them to navigate in conditions of uncertainty. Commenting on his early intuitions in a conversation with Matossian many decades after his first experiments, he updated and consolidated his perspective to conform with modern neurosciences: 'Our attention is unable to follow all the various events, so instead we form a general impression. That's simply how our brain reacts to mass phenomena – there's no question of scientific computations. Our brain does a kind of statistical analysis! Again, the solution is probability theory' (78). In this late interview, Xenakis manages to articulate how stochastic music could be actually considered a universalist, transcultural project. When tonal and, to even larger extent, serial music demands a trained ear, accustomed to certain pitch hierarchies determining the linear succession of sounds, then stochastic music speaks to the pre-cultural and thus pre-individual capacities of the human brain; it is addressed to the uncultured mind, performing its tasks in the realm of *zoe*.

However, if the members of the audience are treated in the most egalitarian way imaginable, stochastic music puts the live performers in a rather awkward position. As the probabilistically modelled transformations of sound clouds in *Pithoprakta* do not generate coherent chains of instructions for the individual performers, no repeating note-to-note relationships exist (see Hill 1975). The composer puts the performers of his music in the uncomfortable position of having to repeat gestures, often at very rapid tempo, which more often than not possess no internal logic. Without necessarily implying that Xenakis's music is any way machinic, the performers themselves can be compared to automatons generating sounds which cannot be comprehended or interpreted in any way. Xenakis did not hide the fact that he sought to assign the performer the role of a technical instrument. For example, concerning the unusual arrangement of the orchestra in the concert hall during the performance of *Terrêtektorh*, he explained:

> The scattering of the musicians brings in a radically new kinetic conception of music which no modern electro-acoustical means could match. (. . .) The speeds and accelerations of the movement of the sounds will be realized, and new and powerful functions will be able to be made use of, such as logarithmic or Archimedean spirals, in-time and geometrically. Ordered or disordered sonorous masses, rolling one against the other like waves . . . etc., will be possible. *Terrêtektorh* is thus a 'Sonotron': an accelerator of sonorous particles, a disintegrator of sonorous masses, a synthesizer. (236–7)

To achieve this 'radically new kinetic conception of music', during live performances of *Terrêtektorh* musicians were dispersed stochastically in a circular space and the audience was seated on folding chairs among them. Xenakis called this piece an apparatus: a machine made of individual organisms that emitted sound waves when orchestrated (synchronised) by the conductor. Stating that a musical piece is a sound generator implies that performers are technical elements (cogs) that mechanically execute instructions. During live performances of Xenakis's early orchestral music, the individuality of the musician as a bios (individual life) is effectively abolished. Like Cage's randomised music, the performer's inability to meaningfully situate their actions – which mimic the behaviour of an insect in a swarm – within a larger framework seem like a dehumanising practice. The performer in *Pithoprakta* must surrender to this senseless discipline, sacrificing their individuality to the totality of the cloud. As I already pointed out, the performer resembles a gas particle, altering its states along with the transformation of the cloud, pushed and pulled by forces that are beyond comprehension from its individual point of view. The identity and coherence of the piece becomes apparent only to an external listener who hears the totality of sounds being produced, or to the composer himself, who knows the algorithm used for generating the notes.[7] Alex Ross describes this

---

[7] There is also the hypothetical possibility of separating the parts from the whole in the reception – a persistent listener can focus their attention on the actions of an individual performer. It is hard to imagine that listening to these sounds, which seem quite random without the context of the whole composition, would be a pleasant or even interesting endeavour. This paradox is easily illustrated by the following analogy: looking at a tree, I can focus my eyes on its entirety, and then on a single leaf. Taking in the whole thing does not entirely blur its details.

state of bewilderment as follows: 'At the height of this meticulously planned bedlam, the listener is incapable of perceiving what any one instrument is doing; only the sum of the actions is apparent' (300).

It seems that from a socio-aesthetic point of view – considering the rules and mechanisms governing a live musical performance – there appears to be little difference between the probabilistic aesthetics of Xenakis and those of John Cage, which I analysed in the previous chapter. In both cases, it is the audience that is given freedom to play with 'the apparatus' of musical performance – to walk around the space, direct their attention freely, etc. – whereas the performers are required to play by the strict rules of composition, demanding nothing less than absolute submission to subjectively nonsensical notation. In both cases, however different they might be in their technical details and their grand philosophical framework, the listeners' enjoyment of the freedom of randomness depends on the performers and artists suppressing individual expression: by randomness generators in Cage, and mathematical procedures in Xenakis. For both artists, harnessing chance for aesthetic purposes hinges upon reining in individual expression.

Interestingly, we might reach a similar conclusion when we look at how Xenakis's music is structured internally, how the particular instructions relate to the whole of a composition. This is exactly how Theodor Adorno advised we look at music: as a material fossil that registers transformations of the social order. He argued that all music – from great orchestral pieces to popular tunes – reflects some truth about social relations at a given historical moment, because every piece of music establishes a system of structural relationships between the individual elements and the whole (Adorno 1976: 55–70). The musical 'individuals' can be identified variously with such elements of musical composition, like the part for a concert instrument, the individual tone, or the motif, which are then regarded in relation to the totality of the musical structure (for a detailed analysis see: Witkin 1998: 28–49). For Adorno, the paradigmatic musical genre of the modern, bourgeois culture is the sonata, which finds its equivalents in other artistic media: in literature in the novel, as analysed by György Lukács (1971), and in middle-class theatre through the dominance of the Absolute Drama (Szondi 1987). In Romantic sonatas, such as those composed by Ludwig van Beethoven, the individual parts of the piece, its motifs, are developed sumptuously on their own, but at the same time, they harmonise organically, and evolve in relation to the musical totality of the composition. Adorno praises this model for the organic equilibrium between parts and the whole, representing a social order in which individuals keep their distinctiveness, yet remain attuned to the social whole. Robert Witkin rightly observes that:

> What Adorno says here of music applies to his idea of individuals in society, too. The idea of an isolated subject or ego is for him a kind of emptiness, a nothingness. It is only as a being, thoroughly mediated by its relations with others, that the individual acquires any kind of mass or substance. (1998: 47)

The sonata's motifs, like the individual in a society, are defined by its relation to other entities and the grand order of the composition.

There are, of course, many issues with this theory, especially with regard to how it was used by Adorno to launch an absurd attack on popular music and on jazz in particular, as a manifestation of the alienated, consumerist society of the industrial age. Adorno also seems unconcerned by the fact that the constituent parts of a sonata do not emerge spontaneously, but are intentionally arranged by a (totalitarian) composer and executed under a guidance of the conductor. In contrast, the notes generated by Xenakis using Gaussian statistical distribution are equally authored by the composer, who assumes certain values to determine the identity of the cloud, and to the mathematical formula, which dictates the various pitches to be played by the musicians. Putting issues of the authorship and intentionality aside, however, Xenakis's stochastic compositions also project a certain image of a social order, very much tied to the Dionysian ideal of the egalitarian society, in which randomness prevails over organisation and, simultaneously, to the industrial society in which the technological environment becomes a crucial and indispensable frame of reference for existence. Again, interestingly enough, wilfully probabilistic music in the form of a finite piece to be played in bourgeois institution seems to only find its true expression in early rave culture, which captured ecstatic, Dionysian social energies within the post-industrial (and post-traumatic) framework of abandoned factories, empty warehouses and electronically produced soundscapes fractured by repetitive and punishingly loud beats.

Bibliography

Adams, Christopher. 2013. 'Historiographical Perspectives on 1940s Futurism'. *Journal of Modern Italian Studies* 18 (4): 419–44.

Adcock, Craig. 1984. 'Conventionalism in Henri Poincaré and Marcel Duchamp'. *Art Journal* 44 (3): 249–58.

Adorno, Theodor W. 1976. *Introduction to Sociology of Music.* New York: The Seabury Press.

Alvarado, Carlos. 2008. 'Note on Charles Richet's "La Suggestion Mentale et le Calcul des Probabilités" (1884)'. *Journal of Scientific Exploration* 22 (4): 543–8.

Arsenault, Linda. 2000. *An Introduction to Iannis Xenakis's Stochastic Music: Four Algorithmic Analyses.* A thesis submitted in conformity with the requirements for the degree of Doctor of Philosophy Graduate Department of Music University of Toronto. https://tspace.library.utoronto.ca/handle/1807/13842.

Ashton, Dore. 1966. 'Interview with Marcel Duchamp'. *Studio International* 171 (878): 244–6.

Auslander, Philip. 2011. *Liveness: Performance in a Mediatized Culture.* London: Routledge.

Austin, John Langshaw. 1962. *How to Do Things with Words.* Oxford: Clarendon Press.

Aziz, Ramy Karam. 2009. 'A Hundred-year-old Insight into the Gut Microbiome!' *Gut Pathog* 1 (1): 21.

Bachelard, Gaston. 1984. *The New Scientific Spirit.* Boston, MA: Beacon Press.

Bachelard, Gaston. 2016. 'Surrationalism'. In Zbigniew Kotowicz, *Gaston Bachelard: A Philosophy of the Surreal.* Edinburgh: Edinburgh University Press: 77–81.

Bain, Alexander. 1902. *The Senses and the Intellect.* New York: D. Appleton and Co.

Baise, Arnold. 2020. 'The Objective-subjective Dichotomy and its Use in Describing Probability'. *Interdisciplinary Science Reviews* 45 (2): 174–85.

Ball, Philip. 2008. 'Quantum Weirdness and Surrealism'. *Nature* 453 (7198): 983–4.

Ball, Philip. 2018. *Beyond Weird: Why Everything You Thought You Knew About Quantum Physics is Different.* Chicago: University of Chicago Press.

Ball, Philip. 2019. 'Mysterious Quantum Rule Reconstructed From Scratch'. *Quanta Magazine.* https://www.quantamagazine.org/the-born-rule-has-been-derived-from-simple-physical-principles-20190213/.

Barthes, Roland. 1982. 'The Death of the Author'. In *A Barthes Reader*, ed. Susan Sontag. New York: Hill and Wang.

Beck, John, and Ryan Bishop. 2020. *Technocrats of the Imagination.* London: Duke University Press.

Beck, Ulrich. 1992. *Risk Society: Towards a New Modernity.* London: Sage Publications.

Becker, Annette. 2000. 'The Avant-Garde, Madness and the Great War'. *Journal of Contemporary History* 35 (1): 71–84.

Bell, John S. 1966. 'On the Problem of Hidden Variables in Quantum Mechanics'. *Rev. Mod. Phys.* 38 (July): 447–52.

Benedetta. [1927] 2009. 'Futurist Sensibility'. In *Futurism: An Anthology*, ed. Lawrence Rainey, Christine Poggi and Laura Wittman, 279–81. New Haven, CT and London: Yale University Press.

Berghaus, Günter. 2009. 'Futurism and the Technological Imagination Poised between Machine Cult and Machine Angst'. In *Futurism and Technological Imagination*, ed. Günter Berghaus. Amsterdam: Rodopi.

Bernoulli, Jakob. [1713] 2005. *The Art of Conjecturing, together with Letter to a Friend on Sets in Court Tennis.* Baltimore: Johns Hopkins University Press.

Bernstein, David W., and Christopher Hatch. 2001. *Writings through John Cage's Music, Poetry, and Art.* Chicago: University of Chicago Press.

Birkhoff, George David. 2013. *Aesthetic Measure.* Cambridge, MA: Harvard University Press.

Bishop, Ryan, and John Phillips. 2010. *Modernist Avant-garde Aesthetics and Contemporary Military Technology: Technicities of Perception.* Edinburgh: Edinburgh University Press.

Bittermann, Henry J. 1940. 'Adam Smith's Empiricism and the Law of Nature: I'. *Journal of Political Economy* 48 (4): 487–520.

Boccioni, Umberto. 1914. 'Il cerchio non si chiude!' In *Lacerba* 2 (5): 67–9.

Boccioni, Umberto. [1913] (2009). 'The Plastic Foundations of Futurist Sculpture and Painting'. In *Futurism: An Anthology*, ed. Lawrence Rainey, Christine Poggi and Laura Wittman, 139–42. New Haven, CT and London: Yale University Press.

Boccioni, Umberto. [1914] (2009). 'ABSOLUTE MOTION + RELATIVE MOTION = DYNAMISM'. In *Futurism: An Anthology*, ed. Lawrence Rainey et al.

Boccioni, Umberto, Carlo Carrà, Luigi Russolo, Giacomo Balla and Gino Severini. [1910] (2009). 'Manifesto of the Futurist Painters'. In *Futurism: An Anthology*, ed. Lawrence Rainey et al.

Boccioni, Umberto, Carlo D. Carrà, Luigi Russolo, Giacomo Balla and Gino Severini. [1910] (2016). 'Technical Manifesto of Futurist Painting'. In *Futurist Painting Sculpture: (Plastic Dynamism)*, ed. Maria Elena Versari. Los Angeles, CA: Getty Research Institute.

Boccioni, Umberto, Carlo D. Carrà, Luigi Russolo, Giacomo Balla and Gino Severini. [1912] (2016). 'Preface to the Catalogue of the First Exhibition of Futurist Painting'. In *Futurist Painting Sculpture: (Plastic Dynamism)*, ed. Maria Elena Versari.

Boccioni, Umberto. 2016. *Futurist Painting Sculpture: (Plastic Dynamism)*, ed. Maria Elena Versari. Los Angeles: Getty Research Institute.

Bohr, Niels. 1963. *Essays on Atomic Physics and Human Knowledge: 1958–1962*. London: Interscience Publishers.

Bohr, Niels. 1985. 'The Bohr-Einstein Dialogue'. In *Niels Bohr: A Centenary Volume*, ed. Anthony Philip French and P. J. Kennedy, 121–40. Cambridge, MA: Harvard University Press.

Bohr, Niels. 1987. *The Philosophical Writings of Niels Bohr*. Woodbridge, CT: Ox Bow Press.

Breton, André. [1924] 1969. 'Manifesto of Surrealism'. In *Manifestoes of Surrealism*. Ann Arbor: University of Michigan Press.

Breton, André. [1925] 1969. 'A Letter to Seers'. In *Manifestoes of Surrealism*.

Breton, André. [1929] 1969. 'Preface for a Reprint of the Manifesto'. In *Manifestoes of Surrealism*.

Breton, André. [1930] 1969. 'Second Manifesto of Surrealism'. In *Manifestoes of Surrealism*.

Breton, André. [1934] 1978. 'What is Surrealism?' In *What is Surrealism? Selected Writings*, ed. Franklin Rosemont. London: Pluto Press.

Breton, André. [1934] 1990. *Communicating Vessels*. Lincoln: University of Nebraska Press.

Breton, André. [1936] 1972. 'The Crisis of the Object'. In *Surrealism and Painting*. New York: Harper & Row.

Breton, André. 1987. *Mad Love*. Lincoln: University of Nebraska Press.

Breton, André. 1996. *The Lost Steps*. Lincoln: University of Nebraska Press.

Breton, André, and Philippe Soupault. [1920] 1985. *The Magnetic Fields*. London: Atlas Press.

Brockman, John. 2014. *By the Late John Brockman*. New York: Harper Perennial.

Broglie de, Louis. 1953. *The Revolution in Physics*. New York: The Noonday Press.

Broude, Norma. 1974. 'New Light on Seurat's "Dot": Its Relation to Photo-Mechanical Color Printing in France in the 1880s'. *The Art Bulletin* 56 (4): 581–9.

Bub, Jeffrey. 2010. 'Von Neumann's "No Hidden Variables" Proof: A Re-Appraisal'. *arXiv*:1006.0499.

Burns, Russell W. 2004. *Communications: An International History of the Formative Years*. London: The Institution of Engineering and Technology.

Byrne, Edmund. 1968. *Probability and Opinion: A Study in the Medieval Presuppositions of Post-Medieval Theories of Probability*. The Hague: Nijhoff.

Cabanne, Pierre. 1979. *Dialogues with Marcel Duchamp*. Boston, MA: Da Capo Press.

Cage, John. 1961. *Silence*. Middletown, CT: Wesleyan University Press.

Cage, John. 1967. *A Year from Monday: New Lectures and Writings by John Cage*. Middletown, CT: Wesleyan University Press.

Cage, John. 1981. *Empty Words: Writings '73–'78 by John Cage*. Middletown, CT: Wesleyan University Press.

Cage, John. n.d. Official website. *John Cage Complete Works: Variations VII*. https://johncage.org/pp/John-Cage-Work-Detail.cfm?work_ID=272.

Cage, John, Michael Kirby and Richard Schechner. 1965. 'An Interview with John Cage'. *The Tulane Drama Review* 10 (2): 50–72.

Caldwell, Bruce. 2004. 'Some Reflections on F. A. Hayek's *The Sensory Order*'. *Journal of Bioeconomics* 6 (3): 239–54.

Cannon, Edmund Stuart, and Ian Tonks. 2008. *Annuity Markets*. Oxford: Oxford University Press.

Caws, Mary Ann. 1964. 'The "réalisme ouvert" of Bachelard and Breton'. *The French Review* 37 (3): 302–11.

Caws, Mary Ann. 1990. 'Introduction'. In André Breton, *Communicating Vessels*, vii–xiii.

Chemero, Anthony. 2009. *Radical Embodied Cognitive Science*. Cambridge, MA: MIT Press.

Chessa, Luciano. 2012. *Luigi Russolo, Futurist: Noise, Visual Arts, and the Occult*. Berkeley: University of California Press.

Chipp, Herschel Browning, Peter Selz and Joshua C. Taylor. 1968. *Theories of Modern Art: A Source Book by Artists and Critics*. Berkeley: University of California Press.

Chirico de, Giorgio. 2001. 'From The Memoirs of Giorgio de Chirico'. In *Surrealist Painters and Poets*, ed. Mary Ann Caws. Cambridge, MA: MIT Press.

Clark, Andy. 2016. *Surfing Uncertainty: Prediction, Action, and the Embodied Mind*. Oxford: Oxford University Press.

Clark, T. J. 1984. *The Painting of Modern Life: Paris in the Art of Manet and his Followers*. Princeton, NJ: Princeton University Press.

Clarke, Bruce. 2002. 'From Thermodynamics to Virtuality'. In *From Energy to Information*, ed. Bruce Clarke and Linda Dalrymple Henderson, 17–34. Stanford, CA: Stanford University Press.

Clausewitz, Carl von. [1832] 1984. *On War*. Princeton, NJ: Princeton University Press.

Coen, Ester. 1988. *Boccioni*. New York: The Metropolitan Museum of Modern Art.

Corra, Bruno, and Emilio Settimelli. [1914] 2009. 'Weights, Measures, and Prices of Artistic Genius'. In *Futurism: An Anthology*, ed. Lawrence Rainey, Christine Poggi and Laura Wittman, 181–6. New Haven, CT and London: Yale University Press.

Cox, Christoph. 2018. *Sonic Flux: Sound, Art, and Metaphysics*. Chicago: University of Chicago Press.

Crary, Jonathan. 2006. *Techniques of the Observer: On Vision and Modernity in the Nineteenth Century*. Cambridge, MA: MIT Press.

Dalí, Salvador. 1997a. 'The Tragic Myth of Millet's *L'Angélus*'. In *The Collected Writings of Salvador Dalí*, 279–334. Cambridge: Cambridge University Press.

Dalí, Salvador. 1997b. 'The Conquest of the Irrational'. In *The Collected Writings of Salvador Dalí*, 256–61.

Dalí, Salvador. [1928] 1997c. 'The New Limits of Painting'. In *The Collected Writings of Salvador Dalí*, 79–93.

Dalí, Salvador. 2001. 'The Stinking Ass'. In *Surrealist Painters and Poets*, ed. Mary Ann Caws, 179–83. Cambridge, MA: MIT Press.

Dalí, Salvador. 2013. *The Secret Life of Salvador Dalí*. Mineola, NY: Dover Publications.

Damisch, Hubert. 1979. 'The Duchamp Defense'. *October* 10 (Autumn): 5–28.

Damisch, Hubert. 2007. 'Polka Dots and Moonbeams'. In *Georges Seurat: The Drawings*, ed. Jodi Hauptman, and Karl Buchberg, 118–23. New York: Museum of Modern Art.

Daston, Lorraine. 1979. 'D'Alembert's Critique of Probability Theory'. *Historia Mathematica* 6 (3): 259–79.

Daston, Lorraine, and Peter Galison. 2007. *Objectivity*. New York: Zone Books.

David, Florence Nightingale. 1962. *Games, Gods and Gambling: The Origins and History of Probability and Statistical Ideas from the Earliest Times to the Newtonian era*. New York: Hafner Pub. Co.

Da Vinci, Leonardo. 2008. *Notebooks*. Ed. Irma A. Richter, Martin Kemp and Thereza Wells. Oxford: Oxford University Press.

Davis, Frederick B. 1973. 'Three Letters from Sigmund Freud to André Breton'. *Journal of the American Psychoanalytic Association* 21 (4): 127–34.

Deak, Frantisek. 1991. 'Kaloprosopia: The Art of Personality. The Theatricalization of Discourse in Avant-Garde Theatre'. *Performing Arts Journal* 13 (2): 6–21.

Demos, T. J. 2007. *The Exiles of Marcel Duchamp*. Boston, MA: MIT Press.

Desnos, Robert. 2017. *Surrealist, Lover, Resistant*. Nanholme Mill: Arc Publications.

Díaz Eva. 2015. *The Experimenters: Chance and Design at Black Mountain College*. Chicago: University of Chicago Press.

Di Bona, Elvira, and Stefano Ercolino. 2019. 'Musil in a Loop: The Other Condition and the Extended Mind'. *Rivista di estetica* 70: 49–59. https://journals.openedition.org/estetica/5067.

Dietz, Klaus, and J. A. P. Heesterbeek. 2000. 'Bernoulli Was Ahead of Modern Epidemiology'. *Nature* 408 (6812): 513–4.

Dolar, Mladen. 2013. 'Tyche, Clinamen, Den'. *Continental Philosophy Review* 46 (2): 223–39.

Donnelly, Kevin. 2015. *Adolphe Quetelet, Social Physics and the Average Men of Science, 1796–1874*. Pittsburgh: University of Pittsburgh Press.

Dorn, Walter L. 1972. 'The Prussian Bureaucracy in the Eighteenth Century'. In Paret Peter, *Frederick the Great: A Profile*. London: Macmillan.

Drucker, Johanna. 2014. 'Diagrammatic and Stochastic Writing and Poetics'. *The Iowa Review* 44 (3), 122–32.

Duchamp, Marcel. 1971. *Dialogues with Marcel Duchamp: The Documents of 20th Century Art*. London: Thames and Hudson.

Duchamp, Marcel. 1975. *The Essential Writings of Marcel Duchamp*, ed. Michel Sanouillet and Elmer Peterson. London: Thames and Hudson.

Duchamp, Marcel. 1982. 'Affectueusement, Marcel: Ten Letters from Marcel Duchamp to Suzanne Duchamp and Jean Crotti'. *Archives of American Art Journal* 22 (4): 2–19.

Duffie, Bruce. 1997. *Composer Iannis Xenakis: A Conversation with Bruce Duffie*. 25 March. http://www.bruceduffie.com/xenakis.html.

Durozoi, Gérard, and Bernard Lecherbonnier. 1972. *Le surréalisme: théories, thèmes, techniques*. Paris: Libraire Larousse.

Duve de, Thierry. 1996. *Kant After Duchamp*. Cambridge, MA: MIT Press.

Eburne, Jonathan P. 2015. 'Approximate Life: The Cybernetic Adventures of Monsieur Wzz . . .' In *Surrealism, Science Fiction and Comics*, ed. Gavin Parkinson, 62–81. Liverpool: Liverpool University Press.

Einstein, Albert. 1905. 'Über die von der molekularkinetischen Theorie der Wärme geforderte Bewegung von in ruhenden Flüssigkeiten suspendierten Teilchen'. *Annalen der Physik* 322 (8), 549–60.

Eliot, T. S. 2014. 'Tradition and the Individual Talent'. In *The Complete Prose of T. S. Eliot*, 105–14. Baltimore: Johns Hopkins University Press.

Elk, Michiel Van, and André Aleman. 2017. 'Brain Mechanisms in Religion and Spirituality: An Integrative Predictive Processing Framework'. *Neuroscience and Biobehavioral Reviews* 73: 359–78.

Ellul, Jaques. 2001. 'Remarks on Technology and Art'. *Bulletin of Science, Technology & Society* 21 (1): 26–37.

Farmelo, Graham. 2009. *The Strangest Man: The Hidden Life of Paul Dirac, Mystic of the Atom*. New York: Basic Books.

Fenyvesi, Kristóf. 2014. 'Dionysian Biopolitics: Karl Kerényi's Concept of Indestructible Life'. *Comparative Philosophy* 5 (2): 45–68.

Feynman, Richard P., Albert R. Hibbs and Daniel F. Styer. 2005. *Quantum Mechanics and Path Integrals*. Mineola, NY: Dover Publications.

Filipovic, Elena. 2016. *The Apparently Marginal Activities of Marcel Duchamp*. New York: MIT Press.

Finetti de, Bruno. 1989. 'Probabilism'. *Erkenntnis* 31 (2–3): 169–223.

Fink, Bruce. 1999. *A Clinical l Introduction to Lacanian Psychoanalysis: Theory and Technique*. Cambridge, MA: Harvard University Press.

Finkelstein, Haim. 1983. 'Salvador Dalí: Double and Multiple Images'. *American Imago* 40 (4): 311–35.

Fischer-Lichte, Erika. 2008. *The Transformative Power of Performance*. London: Routledge.

Fisher, Ronald A. 1935. *The Design of Experiments*. Edinburgh: Oliver & Boyd.

Fleuret, Maurice (eds). 1981. *Regards sur Xenakis*. Paris: Stock-Musique.

Florida, Richard L. 2002. *The Rise of the Creative Class*. New York: Basic Books.

Flusser, Vilém. 1984. *Towards a Philosophy of Photography*. Göttingen: European Photography.

Flusser, Vilém. 1989. 'Von linearen Entscheidungen zu synthetischen Projektionen'. *Gdi impuls* 4: 17–27.

Flusser, Vilém. 2011. *Into the Universe of Technical Images*. Minneapolis: University of Minnesota Press.

Flusser, Vilém. 2013. *Post-history*. Minneapolis: Univocal Publishing.

Foster, Hal. 1996. 'Obscene, Abject, Traumatic'. *October* 78 (Autumn): 107–24.

Foster, Hal. 1997. *Convulsive Beauty*. Cambridge, MA: MIT Press.

Foucault, Michel. 2001. *Madness and Civilization*. London: Routledge.

Foucault, Michel. 2007. *Security, Territory, Population: Lectures at the College De France, 1977–78*, ed. Michel Senellart. London: Palgrave Macmillan.

Friston, Karl. 2010. 'The Free-energy Principle: A Unified Brain Theory?' *Nature Reviews Neuroscience* 11 (2): 127–38.

Fuller, Buckminster R. 1981. *Critical Path*. New York: St. Martin's Press.

Gabor, Dennis. 1946. 'Theory of Communication'. *The Journal of the Institution of Electrical Engineers* 93 (3): 429–57.

Galison, Peter. 2002. 'Images Scatter into Data, Data Gather into Images'. In *Iconoclash: Beyond the Image Wars in Science, Religion and Art*, ed. Bruno Latour and Peter Weibel, 301–21. Cambridge, MA: MIT Press.

Galovotti, Maria Carla. 2008. 'De Finetti's Philosophy of Probability'. In Bruno de Finetti, *Philosophical Lectures on Probability*, xv–xxii. Berlin: Springer.

Gamboni, Dario. 1999. '"Fabrication of Accidents": Factura and Chance in Nineteenth-Century Art'. *RES: Anthropology and Aesthetics* 36 (Autumn): 205–25.

Getsy, David. 2011. *From Diversion to Subversion: Games, Play, and Twentieth-Century Art*. University Park: Penn State University Press.

Giddens, Anthony. 1990. *Consequences of Modernity*. Cambridge: Polity Press.

Gigerenzer, Gerd. 1987. 'Probabilistic Thinking and the Fight against Subjectivity'. In *The Probabilistic Revolution: vol. 2. Ideas in the Sciences*, ed. Lorenz Kruger, Lorraine J. Daston and Michael Heidelberger, 11–33. Cambridge, MA: The MIT Press.

Gigerenzer, Gerd, Zeno Swijtink, Theodore Porter, Lorraine Daston, John Beatty and Lorenz Kruger. 1989. *The Empire of Chance: How Probability Changed Science and Everyday Life*. Cambridge: Cambridge University Press.

Gigerenzer, Gerd, and David J. Murray. 2015. *Cognition as Intuitive Statistics*. New York: Psychology Press.

Gill, Miranda. 2009. *Eccentricity and the Cultural Imagination in Nineteenth-Century Paris*. New York: Oxford University Press.

Gillispie, C. C. 1997. *Pierre-Simon Laplace. A Life in Exact Science*. Princeton, NJ: Princeton University Press.

Goetzmann, William N. 1995. 'The Informational Efficiency of the Art Market'. *Managerial Finance* 21 (6): 25–34.

Griffiths, Paul. 2001. 'Iannis Xenakis, Composer Who Built Music on Mathematics, is Dead at 78'. *New York Times*, 5 February. https://www.nytimes.com/2001/02/05/arts/iannis-xenakis-composer-who-built-music-on-mathematics-is-dead-at-78.html.

Grubbs, David. 2014. *Records Ruin The Landscape: John Cage, the Sixties, and Sound Recording*. Durham, NC: Duke University Press.

Hacking, Ian. 1988. 'Telepathy: Origins of Randomization in Experimental Design'. *Isis* 79 (3): 427–51.

Hacking, Ian. 1990. *The Taming of Chance*. Cambridge: Cambridge University Press.

Hacking, Ian. 2006. *The Emergence of Probability*. Cambridge University Press. eBook.

Hamouda, Omar F., and Robin Rowley. 1996. *Probability in Economics*. London: Routledge.

Harley, James. 2004. *Xenakis: His Life in Music*. London: Routledge.

Harrison, Thomas. 2014. 'The Essayistic Novel and Mode of Life: Robert Musil's The Man without Qualities'. *Republic of Letters* 4. https://arcade.stanford.edu/rofl/essayistic-novel-and-mode-life-robert-musils-man-without-qualities.

Hartkamp, Mirjam, and Ian M. Thornton. 2017. 'Meditation, Cognitive Flexibility and Well-Being'. *Journal of Cognitive Enhancement* 1 (2): 182–96.

Haswell, Russell, and Florian Hecker. 2007. 'Blackest Ever Black'. In *Collapse* III, ed. Robin Mackay. Cambridge, MA: MIT Press.

Hayek, F. A. 1952. *The Sensory Order*. Chicago: University of Chicago Press.

Helmholtz, Hermann von. 1995. *Science and Culture: Popular and Philosophical Essays*. Chicago: University of Chicago Press.

Henderson, Linda Dalrymple. 2002. 'Vibratory Modernism: Boccioni, Kupka, and the Ether of Space'. In *From Energy to Information*, ed. Bruce Clarke and Linda Darlymple Henderson. Stanford, CA: Stanford University Press.

Henderson, Linda Darlymple. 2005. *Duchamp in Context: Science and Technology in the Large Glass and Related Works*. Princeton, NJ: Princeton University Press.

Hill, Peter. 1975. 'Xenakis and the Performer'. *Tempo* 112: 17–22.

Horrocks, Roger. 1994. *Masculinity in Crisis*. London: Palgrave Macmillan.

Hutchins, Edwin. 2014. 'The Cultural Ecosystem of Human Cognition'. *Philosophical Psychology* 27 (1): 34–49.

Jaeger, Peter. 2013. *John Cage and Buddhist Ecopoetics*. New York: Bloomsbury Academic.

Jarzębski, Jerzy. 2003. 'Chaos na cenzurowany: późna eseistyka Lem'. *Zagadnienia filozoficzne w nauce* XXXIII: 47–63.

Joselit, David. 1992. 'Marcel Duchamp's "Monte Carlo Bond" Machine'. *October* 59 (Winter): 8–26.

Joseph, Branden Wayne. 2016. *Experimentations: John Cage in Music, Art, and Architecture*. New York: Bloomsbury Academic.

Judovitz, Dalia. 1995. *Unpacking Duchamp: Art in Transit*. Berkeley: University of California Press. http://ark.cdlib.org/ark:/13030/ft3w1005ft.

Kahn, Douglas. 2013. *Earth Sound Earth Signal:Energies and Earth Magnitude in the Arts*. Berkeley: University of California Press.

Kassung, Christian. 2001. *Entropie-Geschichten: Robert Musils 'Der Mann ohne Eigenschaften' im Diskurs der modernen Physik*. Munich: Fink.

Kendall, A. I. 1909. 'Some observations on the study of the intestinal bacteria'. *The Journal of Biological Chemistry* 6: 499–507.

Kendall, M. G. 1956. 'Studies in the History of Probability and Statistics: II. The Beginnings of a Probability Calculus'. *Biometrika* 43 (1–2): 1–14.

Kepes, György. [1944] 1969. *Language of Vision*. Chicago: Paul Theobald & Co.

Kerényi, Karl. 1976. *Dionysos: Archetypal Image of Indestructible Life*. Princeton, NJ: Princeton University Press.

Kerouac Jack. 1976. *On the Road*. New York: Penguin Books.

Kilroy, Robert. 2018. *Marcel Duchamp's* Fountain. *One Hundred Years Later*. London: Palgrave Macmillan.

Kittler, Friedrich A. 1999. *Gramophone, Film, Typewriter*. Stanford, CA: Stanford University Press.

Klein, Martin J. (1966). 'Thermodynamics and Quanta in Planck's Work'. *Physics Today* 19 (11): 294–302.

Klütsch, Christoph. 2012. 'Early Computer Arts in Stuttgart'. In *Mainframe Experimentalism: Early Computing and the Foundation of the Digital Arts*, ed. Hannah Higgins, 68–89. Berkeley: Univeristy of California Press.

Koch, Christof, and Idan Segev. 2000. 'The Role of Single Neurons in Information Processing'. *Nature Neuroscience* 3 (11) Supplement: 1171–7.

Kostelanetz, Richard. 2003. *Conversing with Cage*. London: Routledge.

Kragh, Helge. 1999. *Quantum Generations: A History of Physics in the Twentieth Century*. Princeton, NJ: Princeton University Press.

Kuhn, Thomas S. 1962. *The Structure of Scientific Revolutions*. Chicago: University of Chicago Press.

Kuhn, Thomas. 1978. *Black-Body Theory and the Quantum Discontinuity*. Chicago: University of Chicago Press.

Kundera, Milan. 2020. 'The Total Rejection of Heritage, or Iannis Xenakis'. In *Encounter: Essays*. New York: Harper.

Lacan, Jacques. 1988. *The Seminar of Jacques Lacan*, Book II. Cambridge: Cambridge University Press.

Lachapelle, Sofie. 2005. 'Attempting Science: The Creation and Early Development of the Institut Métapsychique International in Paris, 1919–1931'. *Journal of the History of the Behavioral Science*s 41(1): 1–24.

Laplace, Pierre Simon. 1902. *A Philosophical Essay on Probabilities*. New York: J. Wiley.

League of American Orchestras. 2014. *Orchestra Repertoire Report 2012/13*. http://www.americanorchestras.org/knowledge-research-innovation/orr-survey/orr-current.html.

Le Bon, Gustave. 1909. *The Evolution of Matter*. London: The Walter Scott Publishing Co.

Lee, Pamela. 2020. *Think Tank Aesthetics*. Cambridge, MA: MIT Press.

Leibniz. 1864–8. *Die Werke von Leibniz*. Ed. Onno Klopp, 303–15. 11 vols. Hanover: Klindworth.

Lem, Stanisław. 2013. *Summa Technologiae*. Minneapolis: University of Minnesota Press.

Lettvin, Jerome. 1989. 'Introduction'. In *Collected Works of Warren S. McCulloch*, ed. Rook McCulloch, 7–20. Salinas, CA: Intersystems Publications.

Link, David. 2016. *Archaeology of Algorithmic Artefacts*. Minneapolis: Univocal.

Liu, Lydia H. 2010. 'The Cybernetic Unconscious: Rethinking Lacan, Poe, and French Theory'. *Critical Inquiry* 36 (2): 288–320.

Lorenz, Edward N. 1972. 'Predictability: Does the Flap of a Butterfly's Wings in Brazil Set off a Tornado in Texas?' American Association for the Advancement of Science. https://static.gymportalen.dk/sites/lru.dk/files/lru/132_kap6_lorenz_artikel_the_butterfly_effect.pdf/.

Lovink, Geert. 1994. 'Review of Kittler Draculas Vermächtnis'. *Mediamatic* 8 (1). http://www.mediamatic.nl/magazine/8_1/Lovink=dracula.html.

Lukács, György. 1971. *The Theory of Novel*. Cambridge, MA: MIT Press.

McCulloch, Warren S. 1974. 'Recollections of the Many Sources of Cybernetics'. *ASC Forum* 2: 5–16. https://eco.emergentpublications.com/Article/896f9d9d-5f87-4ccb-96ef-b2277b873830/academic.

McCulloch, Warren S. 1988. 'What Is a Number, That a Man May Know It, and a Man, That He May Know a Number?' In Warren S. McCulloch, *Embodiments of the Mind*, 1–19. Cambridge, MA: MIT Press.

McCulloch, Warren S., and Walter Pitts. [1943] 1990. 'A Logical Calculus of the Ideas Immanent in Nervous Activity'. *Bulletin of Mathematical Biophysics* 52 (1/2): 99–115.

Mackay, Robin, ed. 2014. *Collapse* VIII.

McLuhan, Marshall. 1994. *Understanding Media*. Cambridge, MA: MIT Press.

Mahoney, Michael S. 1990. 'Cybernetics and Information Technology'. In R. C. Olby, G. N. Cantor, J. R. R. Christie and M. J. S. Hodge (eds), 537–53. *Companion to the History of Modern Science*. London: Routledge.

Marinetti, Filippo Tommaso. [1911] 2009. 'Electrical War'. In *Futurism: An Anthology*, ed. Lawrence Rainey, Christine Poggi and Laura Wittman, 98–104. New Haven, CT and London: Yale University Press.

Marinetti, Filippo Tommaso. [1912] 2009. 'Technical Manifesto of Futurist Literature'. In *Futurism: An Anthology*, ed. Lawrence Rainey et al.

Marinetti, Filippo Tommaso. [1913] 2009. 'Destruction of Syntax–Radio Imagination–Words-in-Freedom'. In *Futurism: An Anthology*, ed. Lawrence Rainey et al.

Marinetti, Filippo Tommaso. [1914] 2009. 'Geometrical and Mechanical Splendor and the Numerical Sensibility'. In *Futurism: An Anthology*, edited by Lawrence Rainey et al.

Marinetti, Filippo Tommaso, and Tullio Crali. 1990. 'Illusionismo plastico di guerra e perfezionamento della terra'. In *Manifesti, proclami, interventi, e documenti teorici del futurismo, 1909–1944*, ed. Luciano Caruso, SPES: Firenze.

Marinetti, Filippo Tommaso, and Pino Masnata. [1933] 2009. 'The Radia: Futurist Manifesto'. In *Futurism: An Anthology*, ed. Lawrence Rainey, Christine Poggi and Laura Wittman, 292–4. New Haven, CT and London: Yale University Press.

Marinetti, Filippo Tommaso, Marcello Puma and Pino Masnata. [1941] 2009. 'Qualitative Imaginative Futurist Mathematics'. In *Futurism: An Anthology*, ed. Lawrence Rainey, Christine Poggi and Laura Wittman. New Haven, CT and London: Yale University Press.

Masson, André. 2001. 'Painting is a Wager'. In *Surrealist Painters, and Poets*, ed. Mary Ann Caws, 47–50. Cambridge, MA: MIT Press.

Matossian, Nouritza. 1990. *Xenakis*. London: Kahn & Averill.

Maturana, Humberto, and Francisco Varela. 1980. *Autopoiesis and Cognition: The Realization of the Living*. Dordrecht: Reidel.

Maxwell, James Clerk. 1873. 'Molecules'. *Nature* 8 (September): 437–41. https://victorianweb.org/science/maxwell/molecules.html.

Mayants, Lazar. 1984. *The Enigma of Probability and Physics*. Dordrecht: Springer.

Mehra, Jagdish, and Helmut Rechenberg. 1982. *The Historical Development of Quantum Theory*. New York: Springer-Verlag.

Meillassoux, Quentin. 2012. *The Number and the Siren: A Decipherment of Mallarmé's* Coup De Dés. Fairmouth: Urbanomic.

Meillassoux, Quentin. 2014. 'The *Coup de Dés*, or the Materialist Divinization of the Hypothesis'. *Collapse* VIII: 813–48.

Meyer, Leonard B. 1957. 'Meaning in Music and Information Theory'. *The Journal of Aesthetics and Art Criticism* 15 (4): 412–24.

Meyer, Leonard B. 1994. *Style and Music: Theory, History, and Ideology*. Chicago: Chicago University Press.

Michaels, Walter Benn. 2011. 'Neoliberal Aesthetics: Fried, Rancière and the Form of the Photograph'. *Nonsite.org* issue #1. https://nonsite.org/article/neoliberal-aesthetics-fried-ranciere-and-the-form-of-the-photograph.

Milan, Serge. 2009. 'The "Futurist Sensibility": An Anti-philosophy for the Age of Technology'. In *Futurism and the Technological Imagination*, ed. Günter Berghaus, 63–76. Amsterdam; New York: Rodopi.

Molderings, Herbert. 2010. *Duchamp and the Aesthetics of Chance*. New York: Columbia University Press.

Mondrian, Piet. [1937] 1964. 'Plastic Art and Pure Plastic Art'. In *Plastic Art and Pure Plastic Art*, ed. Robert L. Herbert, 152–65. Englewood Cliffs, NJ: Prentice-Hall.

Monod, Jacques. 1971. *Chance and Necessity: An Essay on the Natural Philosophy of Modern Biology*. New York: Alfred A. Knopf.

Morton, Timothy. 2007. *Ecology without Nature: Rethinking Environmental Aesthetics*. Cambridge, MA: Harvard University Press.

Musil, Robert. 1990. *Precision and Soul: Essays and Addresses*. Chicago: Chicago University Press.

Musil, Robert. 2017. *The Man without Qualities*. London: Pan Macmillan.

Myers, Charles S. 1915. 'A Contribution to the Study of Shell Shock: Being an Account of Three Cases of Loss of Memory, Vision, Smell, and Taste, Admitted into the Duchess of Westminster's War Hospital, Le Touquet'. *Lancet*, 13 February.

Naumann, Francis M. 1987. 'Marcel Duchamp: A Reconciliation of Opposites'. In *Artist of the Century*, ed. Rudolf E. Kuenzli and Francis M. Naumann, 20–40. Cambridge, MA: MIT Press.

Negrestani, Reza. 2018. 'On Toy Aesthetics: Wittgenstein's Pinball Machine'. *Toy Philosophy* (blog), 2 April 2. https://toyphilosophy.com/2018/04/02/on-toy-aesthetics-wittgensteins-pinball-machine-part-1/.

Nelson, James, ed. 1958. *Wisdom – Conversation with the Elder Wise Men of Our Day*. New York: W. W. Norton & Co. ['Regions which are not ruled by time and space'. Edited version of 'A Conversation with Marcel Duchamp', television interview conducted by James Johnson Sweeney, NBC, January 1956. The interview was filmed at the Philadelphia Museum of Art. Reprinted in *The Essential Writings of Marcel Duchamp*, ed. Michel Sanouillet and Elmer Peterson, 127–37. London: Thames and Hudson, 1975.]

Neumann, John von. 1956. 'Probabilistic Logics and the Synthesis of Reliable Organisms from Unreliable Components'. In *Automata Studies* 34. ed. C. E. Shannon and J. McCarthy, 43–98. Princeton, NJ: Princeton University Press.

Neumann, John von. 1996. *Mathematische Grundlagen der Quantenmechanik*. Berlin: Springer.

Niebisch, Arndt. 2012. *Media Parasites in the Early Avant-Garde*. New York: Palgrave Macmillan.

Nietzsche, Friedrich. [1872] 2000. *The Birth of the Tragedy*. Oxford: Oxford University Press.

Nowotny, Helga. 2015. *The Cunning of Uncertainty*. Cambridge: Polity Press. E-book.

Noys, Benjamin. 2014. *Malign Velocities: Accelerationism and Capitalism*. London: Zero Books.

Omnès, Roland. 1999. *Understanding Quantum Mechanics*. Princeton, NJ: Princeton University Press.

Ortega, Pedro A. 2015. 'Subjectivity, Bayesianism, and Causality'. *ArXiv* abs/1407.4139.

Otto, Walter F. 1965. *Dionysus: Myth and Cult*. Bloomington and London: Indiana University Press.

Pagnoni, Giuseppe, and Wendy Hasenkamp. 2015. 'Remembrance of Things to Come: The Predictive Nature of the Mind and Contemplative Practices – Mind & Life Institute Blog'. *Mind & Life Institute*. 27 August. https://www.mindandlife.org/remembrance-things-come-predictive-nature-mind-contemplative-practices/.

Pais, Abraham. 2005. *Subtle Is the Lord: The Science and the Life of Albert Einstein*. Oxford: Oxford University Press.

Parikka, Jussi. 2015. *A Geology of Media*. Minneapolis: University of Minnesota Press.

Parkinson, Gavin. 2008. *Surrealism, Art and Modern Science: Relativity, Quantum Mechanics, Epistemology*. London: Yale University Press.

Pascal, Blaise. [1670] 2006. *Pascal's Pensées*. (1958. New York: E. P. Dutton & Co.) Project Gutenberg, 233. https://www.gutenberg.org/files/18269/18269-h/18269-h.htm#p_233.

Paz, Octavio. 2011. *Marcel Duchamp: Appearance Stripped Bare*. New York: Arcade Publishing. E-book.

Peckham, Morse. 1965. *Man's Rage for Chaos*. New York: Schocken Books.

Perloff, Marjorie, and Charles Junkerman. 1994. *John Cage: Composed in America*. Chicago: University of Chicago Press.

Petrella, Fausto. 2004. 'Mater Materia Prima: Clinical and Critical Remarks'. In *Boccioni's* Materia, ed. Laura Mattioli Rossi. New York: Guggenheim.

Phelan, Peggy. 1993. *Unmarked: The Politics of Performance*. New York: Routledge.

Piccinini, Gualtiero. 2020. *Neurocognitive Mechanisms: Explaining Biological Cognition*. Oxford: Oxford University Press.

Pickering, Andrew. 2011. *The Cybernetic Brain: Sketches of Another Future*. Chicago: University of Chicago Press.

Piekut, Benjamin. 2011. *Experimentalism Otherwise: The New York Avant-garde and Its Limits*. Berkeley: University of California Press.

Piekut, Benjamin. 2013. 'Chance And Certainty: John Cage's Politics Of Nature'. *Cultural Critique* 84 (Spring): 134–63.

Pitts, Walter, and Warren S. McCulloch. 1947. 'How We Know Universals: The Perception of Auditory and Visual Forms'. *Bulletin of Mathematical Biophysics* 9 (3): 127–47. https://doi.org/10.1007/BF02478291.

Plato, Jan von. 1994. *Creating Modern Probability: Its Mathematics, Physics and Philosophy in Historical Perspective*. Cambridge: Cambridge University Press.

Plotnitsky, Arkady. 2012. *Niels Bohr and Complementarity*. New York: Springer.

Poggi, Christine. 2009. *Inventing Futurism: The Art and Politics of Artificial Optimism*. Princeton, NJ: Princeton University Press.

Poggioli, Renato. 1968. *The Theory of the Avant-garde*. Cambridge, MA: The Belknap Press.

Poincaré, Henri. 2014. *The Foundations of Science*. Lancaster, PA: The New Era Printing Company.

Poincaré, Henri. 2018. *Science and Hypothesis*. London: Bloomsbury Academic.

Polizzotti, Mark. 1995. *Revolution of the Mind: The Life of André Breton*. New York: Farrar, Strauss, and Giroux.

Porter, Theodore M. 1981. 'A Statistical Survey of Gases: Maxwell's Social Physics'. *Historical Studies in the Physical Sciences* 12 (1): 77–116.

Pritchett, James. 1993. *The Music of John Cage*. Cambridge: Cambridge University Press,

Putnam, Hilary. 2002. 'The Nature of Mental States'. In David Chalmers, *Philosophy of Mind: Classical and Contemporary Readings*. New York: Oxford University Press.

Quetelet, Adolphe. 1829. *Recherches statistique*. Brussels: Hayez.

Rancière, Jacques. 2004. *The Politics of Aesthetics: The Distribution of the Sensible*. London: Continuum.

Rappaport, Roy A. 1971. 'Ritual, Sanctity, and Cybernetics'. *American Anthropologist* 73 (1): 59–76.

Re, Lucia. 2011. 'Introduction to "A Woman with Three Souls"'. *California Italian Studies* 2 (1): 3–14. http://www.escholarship.org/uc/item/7k625747.

Reichenbach, Hans. 1938. *Experience and Prediction: An Analysis of the Foundations and the Structure of Knowledge*. Chicago: University of Chicago Press.

Reynolds, Simon. 2010. *Generation Ecstasy: Into the World of Techno and Rave Culture*. New York: Routledge.

Richet, Charles. 1884. 'La suggestion mentale et le calcul des probabilités'. *Revue philosophique de la France et de l'e'trangere*, 18, 609–74.

Rosà, Rosa. 2011. 'A Woman with Three Souls', *California Italian Studies* 2 (1): 16–40. http://www.escholarship.org/uc/item/7k625747.

Rosenblatt, Frank. 1958. 'The Perceptron: A Probabilistic Model for Information Storage and Organization in the Brain'. *Psychological Review* 65 (6): 386–408.

Ross, Alex. 2007. *The Rest Is Noise*. New York: Picador.

Roudinesco, Élisabeth. 1990. *Jacques Lacan & Co: A History of Psychoanalysis in France, 1925–1985*. Chicago: Chicago University Press.

Roudinesco, Élisabeth. 1997. *Jacques Lacan*. New York: Columbia University Press.

Rowbottom, Darrell Patrick. 2015. *Probability*. Cambridge: Polity Press.

Russell, Bertrand. 1912. 'The Philosophy of Bergson'. *The Monist* 22 (3): 321–47.

Salsburg, David. 2013. *The Lady Tasting Tea*. New York: Henry Holt.

Salzman, Eric. 1960. 'In and Out the Piano with Cage'. *New York Times*, 14 February.

Savage, Neil. 2019. 'How AI and Neuroscience Drive Each Other Forwards'. *Nature* 571 (7553): 15–17.

Schechner, Richard. 2013. *Performance Studies: An Introduction*. London: Routledge.

Schinckus, Christophe. 2017. 'From Cubist Simultaneity to Quantum Complementarity'. *Foundations of Science* 22 (4): 709–16.

Schlatter, Mark, and Ken Aizawa. 2008. 'Walter Pitts and "A Logical Calculus"'. *Synthese* 162 (2): 235–50.

Schnapp, Jeffrey T. 2012. 'The Statistical Sublime'. In *The History of Futurism: The Precursors, Protagonists, and Legacies*, ed. Geert Buelens, Harald Hendrix and Monica Jansen, 107–28. Lanham, MD: Lexington Books.

Schrödinger, Erwin. 1995. *The Interpretation of Quantum Mechanics*. Woodbridge, CT: Ox Bow Press.

Schuessler, Rudolf. 2019. 'Probability in Medieval and Renaissance Philosophy'. *The Stanford Encyclopedia of Philosophy*, ed. Edward N. Zalta. https://plato.stanford.edu/archives/sum2019/entries/probability-medieval-renaissance/.

Schwartz, Arturo. 1969. *The Complete Works of Marcel Duchamp*. London: Thames and Hudson.

Seigel, Jerrold. 2009. 'Selves without Qualities? Duchamp, Musil, and the History of Selfhood'. In *The Modernist Imagination: Intellectual Theory and Critical Theory*, ed. Warren Breckman, Peter E. Gordon, A. Dirk Moses, Samuel Moyn and Elliot Neaman. New York: Berghahn Books.

Serres, Michel. 1982. *Hermes: Literature, Science, Philosophy*. Baltimore, MA: The Johns Hopkins University Press.

Shannon, Claude E., and Warren Weaver. 1964. *The Mathematical Theory of Communication*. Urbana: University of Illinois Press.

Silverman, Kenneth. 2012. *Begin Again: A Biography of John Cage*. Evanston, IL: Northwestern University Press.

Simmel, Georg. [1903] (1950). 'The Metropolis and Mental Life' in *The Sociology of Georg Simmel*, ed. K. H. Wolff, 409–24. New York: Free Press.

Simmel, Georg. 2004. *The Philosophy of Money*. London: Routledge.

Skinner, Burrhus Frederic. [1931] 1999. *Cumulative Record: Definitive Edition*. Cambridge, MA: B. F. Skinner Foundation.

Snow, C. P. 1959. *The Two Cultures and the Scientific Revolution*. New York: Cambridge University Press.

Stallybrass, Peter, and Allon White. 1986. *The Politics and Poetics of Transgression*. Ithaca, NY: Cornell University Press.

Stanley, Matthew. 2008. 'The Pointsman: Maxwell's Demon, Victorian Free Will, and the Boundaries of Science'. *Journal of the History of Ideas* 69 (3): 467–91.

Steyerl, Hito. 2017. *Duty Free Art*. London: Verso.

Stigler, Stephen. 1990. *The History of Statistics*. Cambridge, MA: Harvard University Press.

Strindberg, August. [1984] 1996. 'The New Arts! Or The Role of Chance in Artistic Creation'. In *Selected Essays*, ed. Michael Robinson, 103–7. Cambridge: Cambridge University Press.

Szondi, Peter. 1987. *The Theory of Modern Drama*. Minneapolis: University of Minnesota Press.

Tatarkiewicz, Władysław. 1980. *A History of Six Ideas: An Essay in Aesthetics*. The Hague: Martinus Nijhoff / PWN / Polish Scientific Publishers.

Tatlin, Vladimir. 1919. 'The Initiative Individual in the Creativity of the Collective'. http://theoria.art-zoo.com/the-initiative-individual-in-the-creativity-of-the-collective-tatlin/.

Temperley, David. 2007. *Music and Probability*. Cambridge, MA: MIT Press.

Terranova, Tiziana. 2004. *Network Culture: Politics for the Information Age*. London: Pluto Press.

Thomas, Cyril, André Didierjean, and Serge Nicolas. 2016. 'Scientific Study of Magic: Binet's Pioneering Approach Based on Observations and Chronophotography'. *The American Journal of Psychology* 129 (3): 313–26.

Tomkins, Calvin. 1972. *The World of Marcel Duchamp*. New York: Time.

Tomkins, Calvin. 2013. *Marcel Duchamp: The Afternoon Interviews*. New York: Badlands Unlimited.

Turner, Charles. 2005. '"Vers une métamusique": Xenakis, Sieve Theory and Cultural Plurality'. In *Definitive Proceedings of the 'International Symposium Iannis Xenakis'* (Athens, May 2005). www.iannis-xenakis.org.

Turner, Fred. 2006. *From Counterculture to Cyberculture*. Chicago: University of Chicago Press.

Tutino, Stefania. 2018. *Uncertainty in Post-Reformation Catholicism: A History of Probabilism*. Oxford: Oxford University Press.

Tzara, Tristan. 2005. *Approximate Man & Other Writings*. Boston, MA: Black Widow Press.

Van de Cruys, Sander. 2014. *To Err and Err, But Less and Less: Predictive Coding and Affective Value in Perception, Art, and Autism*. PhD diss., Faculty of Psychology and Pedagogical Sciences KU Leuven.

Varèse, Edgard. [1933] 2004. 'The Liberation of Sound'. In *Audio Culture*, ed. Christoph Cox and Daniel Werner, 17–21. London: Bloomsbury Academic.

Varga, Bálint András. 1996. *Conversations with Iannis Xenakis*. London: Faber and Faber.

Versari, Maria Elena, ed. 2016. 'Introduction'. In Umberto Boccioni, *Futurist Painting Sculpture: (Plastic Dynamism)*. Los Angeles: Getty Research Institute.

Weaver, Warren. 1949. 'Introduction'. In Claude E. Shannon and Warren Weaver, *The Mathematical Theory of Communication*. Chicago: University of Illinois Press.

Weaver, Warren. 1964. 'Introductory Note on the General Setting of the Analytical Communication Studies'. In Claude E. Shannon, *The Mathematical Theory of Communication*. Urbana: University of Illinois Press.

Weil, Simone. [1941] 2015. 'At the Price of an Infinite Error: The Scientific Image, Ancient and Modern'. *Simone Weil: Late Philosophical Writings*, ed. Eric O. Springsted. Notre Dame, IN: University of Notre Dame Press.

Whyte, Iain Boyd. 2000. 'The Architecture of Futurism'. In *International Futurism in Arts and Literature*, ed. Günter Berghaus, 353–72. Berlin: Walter de Gruyter.

Wilson, Edwin Bidwell. 1923. 'First and Second Laws of Error'. *Journal of the American Statistical Association* 18 (143): 841–51.

Witkin, Robert A. 1998. *Adorno on Music*. London: Routledge.

Wohl, Victoria. 2014. *Probabilities, Hypotheticals, and Counterfactuals in Ancient Greek Thought*. Cambridge: Cambridge University Press.

Wojnowski, Konrad. 2017. 'Telematic Freedom and Information Paradox'. *Flusser Studies* 23 (1).

Wojnowski, Konrad. 2019. 'War and Dissociation: The Case of Futurist Aesthetics'. In *Politics of the Machines Beirut: Art/Conflict*. The International University of Beirut, Lebanon, 11–14 June, ed. Morten Søndergaard, Laura Beloff, Hassan Choubassi and Joe Elias, 15–20. Beirut, Lebanon: Electronic Workshops in Computing (eWiC).

Xenakis, Iannis. 1955. 'La crise de la musique sérielle'. *Gravesaner Blätter* 1 (July): 2–4.

Xenakis, Iannis. 1970. 'Towards a Metamusic'. *Tempo* 93 (Summer): 2–19. http://www.jstor.org/stable/944414.

Xenakis, Iannis. 1987. 'Xenakis on Xenakis'. *Perspectives of New Music* 25 (1/2): 16–63.

Xenakis, Iannis. 1992. *Formalized Music*. New York: Pendragon Press.

Žižek, Slavoj. 2014. *What Does Europe Want?* New York: Columbia University Press.

Zylinska, Joanna. 2017. *Nonhuman Photography*. Cambridge, MA: MIT Press.

Index

Note: page references in *italics* indicate figures; 'n' indicates footnotes.

EU representative:
Easy Access System Europe
Mustamäe tee 50, 10621 Tallinn, Estonia
Gpsr.requests@easproject.com

www.ingramcontent.com/pod-product-compliance
Lightning Source LLC
Chambersburg PA
CBHW051211170526
45166CB00005B/1850

* 9 7 8 1 4 7 4 4 8 8 9 7 6 *